A *New York Times* Editors' Choice

A *Financial Times* Business Book of the Month

"*Faster, Higher, Farther: The Volkswagen Scandal* takes readers through the combination of pressures that produced what may be the biggest corporate scandal ever, detailing the company's personalities and the history behind the saga with fluency and wit."

—*Atlantic*

"Perversely engaging."

—*Barron's*

"Exhaustive."

—Richard Epstein, *Forbes*

"A fantastic book."

—Catherine Wolfram, co-director of the Energy Institute at Haas, University of California, Berkeley

"This book, which races along like Jensen Button, tells the inside story of the Volkswagen scandal. Ewing tells it quite beautifully."

—*Daily Mail*

"Ewing reveals for the first time the true extent of the scandal."

—*The Times* (London)

"A must read."

—*Handelsblatt*

Faster, Higher, Farther

Faster, Higher, Farther

HOW ONE OF THE WORLD'S LARGEST AUTOMAKERS COMMITTED A MASSIVE AND STUNNING FRAUD

JACK EWING

W. W. NORTON & COMPANY
Independent Publishers Since 1923
New York | London

Printed in the United States of America
First published as a Norton paperback 2018

For information about permission to reproduce selections from this book, write to
Permissions, W. W. Norton & Company, Inc., 500 Fifth Avenue, New York, NY 10110

For information about special discounts for bulk purchases, please contact
W. W. Norton Special Sales at specialsales@wwnorton.com or 800-233-4830

Manufacturing by Quad Graphics Fairfield
Book design by Chris Welch Design
Production manager: Julia Druskin

Library of Congress has cataloged the hardcover edition as follows:

Names: Ewing, Jack, 1955– author.
Title: Faster, higher, farther : the Volkswagen scandal / Jack Ewing.
Description: First edition. | New York : W. W . Norton & Company,
Independent Publishers Since 1923, [2017] |
Includes bibliographical references and index.
Identifiers: LCCN 2017012672 | ISBN 9780393254501 (hardcover)
Subjects: LCSH: Volkswagenwerk—History. | Automobile industry and
trade—Germany—History. | Corporations—Corrupt practices—United States. |
Volkswagen automobiles—Motors (Diesel)—Exhaust gas.
Classification: LCC HD9710.G44 E95 2017 | DDC 338.7/6292220943—dc23
LC record available at https://lccn.loc.gov/2017012672

ISBN 978-0-393-35591-8 pbk.

W. W. Norton & Company, Inc.
500 Fifth Avenue, New York, N.Y. 10110
www.wwnorton.com

W. W. Norton & Company Ltd.
15 Carlisle Street, London W1D 3BS

1 2 3 4 5 6 7 8 9 0

To my father, whose lifelong engagement with the environment informs every page of this book.

CONTENTS

BUILDING AN EMPIRE

VOLKSWAGEN'S PATH TO WORLD DOMINATION

1937 The Nazi labor front founds a company to build a "people's car" or Volkswagen. The familiar circular VW logo is designed by Franz Xaver Reimspiess, a motor specialist in Ferdinand Porsche's design bureau.

1965 Volkswagen buys Auto Union from Daimler-Benz.

1969 Volkswagen merges Auto Union with NSU Motorenwerke to form Audi.

1986 Volkswagen acquires SEAT, state-owned Spanish automaker, with which it has already been jointly building cars.

1991 Following the fall of communism in Eastern Europe, Volkswagen acquires Skoda from the Czech government. It becomes the company's budget brand.

1998 In an attempt to move up market, Volkswagen buys British luxury car maker Bentley Motors, Italian sports car maker Lamborghini, and moribund luxury brand Bugatti.

2008 Volkswagen becomes majority shareholder in Swedish truck maker Scania.

2011 Volkswagen acquires a majority in German truck maker MAN.

2012 Volkswagen acquires Porsche, the sports car maker with which it has long cooperated. The deal leaves the Porsche and Piëch families with a majority of Volkswagen's voting shares.

2012 Volkswagen buys Italian motorcycle maker Ducati.

2015 For the first time, Volkswagen sells more vehicles than Toyota and becomes the world's largest carmaker.

THE PORSCHE AND PIËCH FAMILIES AND VOLKSWAGEN

(Members with the greatest influence on Volkswagen history)

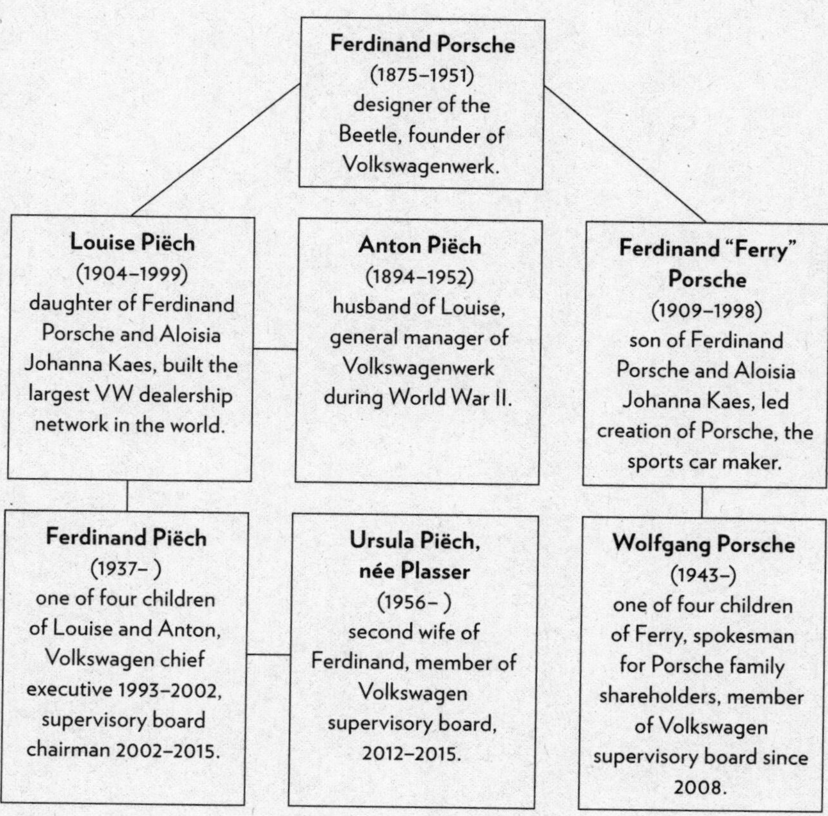

Faster,
Higher,
Farther

CHAPTER 1

Road Trip

THEY WERE A CURIOUS SIGHT, the graduate students from West Virginia University, barreling down California freeways in the spring of 2013. The back end of their car, a Volkswagen Jetta station wagon, sprouted a tangle of pipes and hoses held together with hardware store clamps and brackets. Flexible tubes sucked exhaust from the tailpipes and fed the gas into a mysterious gray box sitting on a slab of plywood in the car's rear cargo area. The box had wires and cables coming out of it. Next to the box, bolted to the plywood, was a Honda portable generator, which stank and made an infernal racket. The students, Hemanth Kappanna from India and Marc Besch from Switzerland, tolerated the noise and the fumes. They had to. The generator was needed to power the whole mess.

People stared. They were questioned by a curious cop. The improvised equipment broke down often. The generator wasn't made to be bumped around so much and needed to be replaced, slowly draining the modest $70,000 grant that West Virginia University had received to fund the research by Kappanna and Besch and another student, named Arvind Thiruvengadam. After one breakdown, Besch and Thiruvengadam spent much of a night in the parking lot of a big-box home improvement store, trying to get the rig to function right again. But the work the students were doing was important—much more important than they could possibly have imagined at the time. They were testing the Jetta's emissions. In particular, they were testing for nitrogen oxides, a family of gases with a wide array of fearsome effects on human health and the environment. Nitrogen oxides cause children to get asthma and provoke attacks in people who already have asthma. They cause chronic bronchitis, cancer, and cardiovascular problems. Excess nitrogen oxides in urban areas have been known to produce spikes in the number of people coming to hospital emergency rooms with heart attacks. Members of the nitrogen oxide family contribute to acid rain and are far more potent, pound for pound, than carbon dioxide as a cause of global warming. Nitrogen oxides also react with sunlight to produce the smog that smothers urban areas, especially Los Angeles, where the students spent much of their time. With its automobile culture, abundant sunlight, and bowl-shaped topography, Los Angeles is an ideal caldron for smog. Thanks largely to nitrogen oxides, LA has the worst air of any city in the United States.

The reason the students were testing a Volkswagen Jetta was that it was one of the few vehicles available in the United States with a diesel motor. They also tested a diesel Volkswagen Passat and a diesel BMW SUV under the supervision of Dan Carder, head of West Virginia University's Center for Alternative Fuels, Engines, and Emissions, or CAFEE, which is famous for its expertise in measuring and analyzing what comes out of a tailpipe. Diesels make more efficient use of fuel than gasoline-powered cars and produce less carbon dioxide. But they also produce far more nitrogen oxides. That's because diesel ignites at higher temperatures than gasoline. The heat turns the inside of a diesel engine into a veritable nitrogen oxide factory, combining nitrogen and oxygen from the atmosphere to form malignant nitrogen oxide molecules.

Volkswagen claimed that the Jetta and Passat were "clean diesels." They were equipped with technology that was supposed to scrub nitrogen oxides out of the exhaust. The German automaker had spent millions of dollars trying to convince Americans that diesels were an environmentally friendly alternative to Toyota's hybrid technology.

That's not what the students from West Virginia University were seeing, though, as they drove around Los Angeles and San Francisco and even up to Seattle. One student was at the wheel while the other sat in the passenger seat with a laptop computer, monitoring the data. Even an expert might have been puzzled by the sight of them. Technology to measure emissions on the road had been around since the 1990s, but it was rarely used for passenger cars. Government regulators preferred to test cars in laboratories, where it was much easier to control all the variables, like barometric pressure or air temperature, that can influence emissions output. The work that the students were doing wasn't exactly revolutionary, but it was unexpected.

The Jetta and Passat emissions were fine when the West Virginia crew tested them on rollers in a specially equipped garage borrowed from the California Air Resources Board, the state's clean air enforcer. But when the students took the Jetta out on the road and hooked up their gear, the Jetta started producing nitrogen oxides in quantities that were off the charts. In fact, the Jetta was producing way more nitrogen oxides than a modern long-haul diesel truck. The Passat was better, but still far above the legal limits. The BMW performed fine, except during a few demanding uphill climbs.

Kappanna couldn't figure it out. He kept expecting the Volkswagen emissions to average out over time at a level somewhere close to the legal limit. That wasn't happening. "Man," Kappanna thought, "this is not controlling very well." He and the others figured there was some kind of mysterious technical problem. Pollution control systems are complex rolling chemistry labs that endeavor to neutralize all the poisons that are a byproduct of modern mobility, not only nitrogen oxides but also other pollutants such as formaldehyde and soot particles. The engineers who design the systems and program the engine computers must manage dozens of variables. It's easy for a clogged valve or a software bug to throw the system out of whack.

Kappanna and the others never suspected anything underhanded on Volkswagen's part. Like almost everyone involved with the automobile industry, they had a lot of respect for German engineering. After all, it was a German, Carl Benz, who filed a patent in 1886 for what is considered the first practical automobile. Ever since, German inventors such as Ferdinand Porsche had been at the forefront of automotive technology. BMW, Mercedes-Benz, and Volkswagen's Audi unit dominated the high end of the auto market. Consumers were willing to pay more for cars simply because they were made in Germany. The whole German economy revolved largely around auto manufacturing. Arguably no one was better at it. The idea that "clean diesel" might be a big lie, and that the lie might be exposed by a handful of underfunded university researchers—they never imagined that in a million years.

CHAPTER 2

The Grandson

IT BEGAN AS A PROPAGANDA EXERCISE. At least since the 1920s, there had been talk in German engineering and political circles about how to build a car that the masses could afford—a people's car, or *Volkswagen*. Part of the motivation was national pride. Even though Carl Benz had invented the automobile, to the chagrin of the Germans it was an American, Henry Ford, who made the new form of transportation affordable. In 1938, there was only one car for every fifty people in Germany. In the United States, thanks in part to Ford, there was one car for every five people.

Hitler seized on the idea of a people's car soon after coming to power in 1933 and turned it into a Nazi prestige project. The car, he decreed, would cost no more than 1,000 reichsmarks, a price chosen because it had a nice ring to it, not because of its achievability. But practicality was never really the point. The Volkswagen was supposed to embody the improvements in living standards the Nazis claimed were just around the corner. And there was an ulterior motive. The car helped justify the autobahns that Hitler was laying down across Germany, and camouflage their real, military purpose. The divided, limited-access highways were designed to allow Wehrmacht convoys to race quickly to German borders.

The Führer wanted a people's car, but the established carmakers like

Daimler-Benz or Adam Opel, which belonged to General Motors, did not. While paying lip service to Hitler's pet project, they were alarmed at the idea of a new competitor subsidized by the state. The automakers' trade association, which was supposed to oversee the project, could not openly oppose it. But it did what it could to undermine the man chosen to design the new car: an engineer and favorite of Hitler's named Ferdinand Porsche. Porsche ran an independent design studio already suffering from financial problems. The association allocated him a meager budget of 200,000 marks to bring the project to completion. That included only 25,000 marks in advance.

But the chiefs of the German auto industry underestimated Porsche's determination and his single-minded passion for engineering. A native of the Sudetenland, which had been part of Austria before World War I and became part of Czechoslovakia afterward, Porsche was already prominent in the fledgling auto industry. He had built a battery-powered car around the turn of the century and during World War I oversaw motorization of Austrian artillery at the Skoda automobile works in what is now the Czech Republic. (Many years later, after the fall of the Berlin Wall, Volkswagen would acquire Skoda.) Between the wars, working mostly as an independent contractor, Porsche designed and built a series of innovative race cars for companies including Daimler-Benz and Auto-Union, which would later become part of Audi. Though he had never earned a university degree and was largely self-taught, Porsche's reputation as an engineer was such that Josef Stalin tried to lure him to the Soviet Union to oversee vehicle construction there.

Hitler invited Porsche to Berlin in March 1934 to discuss the project. They met at the Kaiserhof, the luxury hotel near the Reich Chancellery, where Hitler had lived shortly before taking power. The engineer made a good impression and was formally awarded the project three months later. Though the task was daunting, Porsche was not starting from scratch. He was able to draw heavily on design work his office had performed for previous clients but had never gotten past the prototype stage. The Porsche design bureau in Stuttgart had constructed a test vehicle known as the Porsche Type 32 for a German motorcycle manufacturer that had contemplated getting into the automobile business but ultimately decided against it. The Type 32 had a body design very similar to the one later found on the Volkswagen and some of the same technical features, such as an air-cooled engine mounted in the rear, which eliminated the need for a radiator and saved weight.

Porsche was aware of efforts by established automakers to undercut him, and he made sure to cultivate his relationship with Hitler. On December 29, 1935, for example, Porsche drove a prototype from Stuttgart to Munich in the company of his son, Ferdinand, better known as Ferry, and Anton Piëch, his son-in-law, who was already a member of the Nazi Party. (Piëch is pronounced "pee-echhh," with the "ch" taking a slightly guttural tone.) In Munich, the group met Hitler and gave him a progress report. Hitler was pleased. By keeping open a direct line to Hitler, Porsche protected himself from the other auto manufacturers who were still trying to make sure the Volkswagen never made it to production. The rivals insisted, for instance, that the car not be manufactured until it had completed 50,000 kilometers, or about 30,000 miles, of road testing. That was far more than any other new vehicle had to endure at the time. Porsche's defenders later insisted that he had no love for the Nazis. But he was more than willing to exploit his connection to Hitler when it was necessary to realize his engineering ambitions.

The hostility of the established manufacturers eventually backfired on them. Sensing their lack of enthusiasm, the Deutsche Arbeitsfront, the German Labor Front, took control of the Volkswagen project in 1937. The front was a Nazi-controlled organization that had forcibly absorbed the independent labor unions after Hitler came to power. It had also confiscated the unions' wealth, some of which it now allocated to finance the construction of a factory to build the Volkswagen. The theft of worker property would have important consequences later on. After the war, in return for giving up claims for compensation, Volkswagen employees won a say in the management of the company that was extraordinary even by German standards. Worker representatives on the Volkswagen supervisory board could, for example, veto plans to close a factory. The Nazis had unwittingly laid the groundwork for one of the grandest experiments ever in worker-management cooperation.

As design of the car progressed, thoughts turned to how to mass-produce it. Porsche was by then already in his sixties. Photos from the time show a dour, slightly heavyset man who combed his thinning hair backward and wore a thick mustache. He was capable of monumental temper tantrums when something didn't perform as expected. Porsche had visited America in 1937 and toured the Ford plant, where he even recruited some Ford executives who had German roots. He began planning a factory that would be consciously modeled on Ford mass assembly methods. The site chosen for the new plant was near Fallersleben, a town adjacent to the rail line connecting Hanover and Berlin. Besides having a good railroad link, the site bordered the Mittelland Canal, a man-made waterway that allowed coal for the factory's power plant to be delivered by river barge. In other respects, the factory was in an isolated location, relatively far from any of Germany's major population centers. The location of the city that grew up on the other side of the canal arguably encouraged a provincial, insular view of the world that would prove damaging to Volkswagen decades later. Unlike BMW in Munich or Daimler in Stuttgart, Volkswagen was out in the sticks.

Hitler himself spoke at the ceremony on May 26, 1938, to mark the laying of the Volkswagenwerk foundation. The audience numbered seventy thousand. The car, Hitler announced, would be called the KdF Wagen, short for Kraft durch Freude Wagen—the Strength through Joy Car. Porsche's design closely resembled the vehicle that would later win fame as the Beetle. Given the car's appearance, the nickname was almost inevitable. Because the rear-mounted, air-cooled engine eliminated the need for a radiator grill, Porsche was able to give the KdF Wagen a rounded front end, set off by a similarly shaped roof and prominent fenders designed to minimize wind resistance. Along with the protruding round headlights, the car created the impression of something that was half machine, half bug. The design may not have been completely original, though. Later Porsche was accused of copying elements of a prototype designed for the Czech automaker Tatra by its chief of construction, Hans Ledwinka. The vehicle designed by Ledwinka did indeed have a body shape similar to Porsche's car as well as an air-cooled, rear-mounted motor. Ledwinka protested, but a lawsuit against Porsche was settled by force when German troops invaded Czechoslovakia in March 1939. The price for the KdF Wagen was set at 990 marks, or the equivalent of $396 at the time. Eventually, insurance and a delivery charge added another 250 marks. Germans could pay 5 marks a week on an installment plan that, in theory, entitled them to a car after four and a half years. At the time, the cheapest automobiles on the market cost around 1,700 marks. The car would give average Germans access to modern mobility the same way that the $14 Volksradio, another Nazi project, had given them access to radio broadcasts.

But even at the time some outsiders saw that the KdF Wagen was more slogan than reality and rightly doubted whether Germany could deliver the vehicle in the promised numbers at the advertised price—which in any case was beyond the means of most Germans. Otto D. Tolischus, a *New York Times* correspondent famed for his reporting on prewar Germany, noted skeptically in 1938 that the price of the car kept rising and that the road to mass mobility in Germany was "long and steep." It was Tolischus who first described the car as a "beetle," and the nickname stuck.

Hitler's ambition for Volkswagen was grandiose, bordering on delusional. The factory would produce 1.5 million vehicles a year when it reached full capacity in 1946, the Führer promised. That would be more than Ford. The factory would not only be the biggest auto plant in the world; it would be the biggest factory of any kind. Volkswagen's fixation on bigness for its own sake would survive the war.

By the time Hitler visited the plant again, on June 7, 1939, it had taken on dimensions that fulfilled at least part of his ambition. The complex ran for more than a kilometer along the banks of the Mittelland Canal. The brick façade that faced the waterway was broken by high, narrow rectangular windows, and by brick enclosures that protruded forward at even intervals and housed staircases. The effect was of figures standing at attention. The *Werkshallen*, the factory buildings, had high roofs supported by steel or concrete columns. Rows of angled skylights let in natural light.

But that September, before mass production of the Volkswagen could start, Germany invaded Poland, and the Volkswagenwerk was ordered to concentrate on military production. Output of the KdF Wagen proved to be laughably small in relation to the Nazis' grandiose promises. By the end of the war, the factory had produced just 640 vehicles for civilian use, and all of these went to members of the elite. The 336,000 Germans who had obediently made payments of 5 marks a week toward a car never got their money's worth.

Instead of making Volkswagens, the plant served as a repair facility for Junkers Ju 88 aircraft, a twin-engine dive-bomber that wreaked havoc over Portsmouth and the Isle of Wight during the Battle of Britain. The factory also produced an eclectic collection of military goods, including field ovens, land mines, parts for the V-1 rocket, and *Panzerfäuste*, the handheld antitank weapons best known for their role in the last-ditch attempt by the Nazis to stop the Allied onslaught. The weapons were handed out by the thousands to civilians, including women, children, and the elderly, in the final months of the war.

The Volkswagenwerk also made vehicles, of course, about 66,000 by the end of the war, still a mere fraction of what had been planned. The most important was the so-called Kübelwagen, also designed by Ferdinand Porsche and based on components developed for the KdF Wagen. It was an early example of the adaptability and versatility of Porsche's design for the Volkswagen. The Kübelwagen was in effect a militarized Beetle, the Wehrmacht's equivalent of a Jeep. (In the 1970s, Volkswagen briefly sold a vehicle in the United States called the Thing, which had removable doors and a fold-down windshield and bore a distinct resemblance to the Kübelwagen.) There was even an amphibious version, the Type 166 Schwimmwagen.

Separately, Ferdinand and his son, Ferry, designed an early version of the Tiger Tank that was to be produced by the Krupp armaments concern. The tank was a famous failure. Hitler wanted the Porsches to design a behemoth that would outclass Russian armor the way that Porsche-designed race cars had triumphed on prewar tracks. Germany would have been better off with the panzer equivalent of a Volkswagen. The over-engineered Porsche Tiger was too finicky for the muddy battlefields of the eastern front. Germany struggled throughout the war to build tanks as effective, robust, and simple to maintain as the ones fielded by the Red Army. "It was madness," the historian William Manchester wrote of the tanks designed by Ferdinand and Ferry Porsche. "They belonged in a toy shop, not a weapons forge."

It's quite possible that Ferdinand and Ferry Porsche unintentionally did more for the Allied war effort than they did for the German cause. They squandered precious raw materials on weapons projects that proved impractical on the battlefield. As subsequent decades would show, the Porsche family had a weakness for elegantly engineered mechanical marvels that were too costly or temperamental for the real world. One Porsche-designed tank did prove useful, however. Years later, a surviving example of a Leopard tank designed by the Porsches was used at the Porsche sports car factory in Stuttgart to flatten rejected prototypes of new models, lest they fall into the hands of the competition.

In the summer of 1942, some members of Ferdinand Porsche's family visited the City of the KdF Wagen. Among them was a grandson, Ferdinand Piëch. He was the son of Louise Porsche, Ferdinand Porsche's daughter, and Anton Piëch, who had become a top manager at the Volkswagenwerk. It quickly became evident that young Ferdinand, only five that summer, had inherited not only his grandfather's first name but also his fascination with machines. The grandson had already enjoyed loitering in the Porsche engineering bureau, which had been relocated to Austria, where employees of Ferdinand Porsche Sr. worked on technical projects among rows of drafting tables. Ferdinand Piëch later recalled causing a stir among the adults when he bragged that Germany would soon have rockets that could shoot straight up in the air. The lad had overheard top secret discussions about the V-2, the ballistic missile the Germans later fired at Britain.

At the City of the KdF Wagen, young Ferdinand was allowed to spend the days riding the locomotive that shuttled fuel and raw materials between factory buildings. Left in the care of the locomotive drivers, he helped shovel coal and could see the Junkers bombers lined up for repairs. It was a strange playground for a child not yet old enough to attend school. The Wehrmacht was gearing up for the attack on Stalingrad, which early the next year would end disastrously for Germany and mark a turning point in the war. Long-range British and American bombers had begun attacking German cities. The KdF Wagen factory was not bombed until 1944, but it was a natural target. Moreover, the factory made extensive use of slave labor, including Russian prisoners of war, women from occupied countries such as Poland, and Jews from concentration camps, including Auschwitz. By 1944, two-thirds of the labor at the Volkswagenwerk, some twenty thousand people, was forced.

The Porsches, who lived most of the time in the Alpine idyll of Zell am See, were not an ordinary family. The children began learning the rudiments of how to drive at age five, when they were allowed to operate the stick shift of the family KdF Wagen from the front passenger seat. (The Porsches naturally had one of the few produced.) Shifting in those days was tricky even for adults, to say nothing of a child. The KdF Wagen did not have a synchronized transmission, which meant that shifting involved pressing down the clutch, moving the stick to neutral, letting out the clutch, and then applying the accelerator. The idea was to get the drive shaft and transmission gear spinning at the same rate, so that they could mesh smoothly. Then the driver pressed in the clutch and pushed the stick back into gear. In the Piëch household, the children were expected to time the shifting of the gearstick while the adult operated the clutch and gas pedal. If the child chose the wrong moment to put the car into gear, the transmission would make an agonizing grinding sound. In that case, the child would not be allowed to do any more shifting for a while.

Piëch, a handsome lad with piercing eyes, was probably too young to perceive that many Volkswagen workers were underfed prisoners and de facto slaves. In any event, decades later Piëch recorded only his fascination with the factory and what went on there. During a long walk that summer with his mother, she asked him what he wanted to be when he grew up. "I said," Ferdinand Piëch wrote later, "I would like to someday work in this factory, but not like my father and grandfather at a big desk with paper, but for real, down there, where the workers repaired the airplanes and rode the train, for real, with my hands." A seed had already been planted, a longing that would draw Ferdinand Piëch back decades later and have a profound effect on Volkswagen's history.

Piëch later excused his grandfather's role in the Nazi war machine by saying that Ferdinand Porsche was fearfully naïve about politics, as if that somehow also included being blind to the suffering of forced laborers or being unaware of the regime's brutality. In any case, Porsche had shown that he was clever enough about politics to know how to use his relationship with Hitler when it suited him. It would have been impossible for an adult to ignore the conditions under which the workers toiled. Even the best-treated among them, like the Polish women, were poorly fed and clothed and were denied medicines if they became sick. The worst-treated were the concentration camp inmates, mostly Jews, imported at the request of company managers facing a severe labor shortage and anxious to keep the plant operating at capacity. The inmates were beaten by SS guards, barely fed, and made to work in winter weather in flimsy concentration camp clothing. Deaths were commonplace. True, conditions were better than at Auschwitz, from where many of the Jewish laborers had come, but they were still deplorable.

Perhaps the cruelest aspect of the wartime Volkswagenwerk was the treatment of newborn children. Sara Frenkel, a Jewish woman from Poland who assumed a false identity and worked as a nurse at the Volkswagen factory, described how children born to laborers were taken away soon after birth and kept at a "children's home" in the village of Rühen, northeast of the factory. "The toddlers lay in filth, and it stank of urine and feces. The food was bad and there was not enough water,"

Frenkel recalled later. "The babies were so beautiful, but eventually they all died in Rühen." Historians put the number of children who died in the home from illness and neglect at 365. Hans Körbel, the SS doctor who oversaw the children's facility, was later convicted of war crimes by a British military tribunal and executed in 1947.

Hans Mommsen, author of an exhaustive study of Volkswagen during World War II, concluded that Porsche had "sleepwalked" through the crimes committed at the factory. He was so focused on technical and production goals that he willfully ignored the human cost.

Volkswagen was hardly alone among German corporations in exploiting forced laborers. Many of the companies that are among Germany's largest today, including Daimler, BMW, and Siemens, were guilty of similar practices. But Volkswagen used an unusually large number of slave laborers and was a pioneer in exploiting them. Lacking an established workforce when the war broke out, Volkswagen sought to make up the shortfall with foreigners, most of whom were there under duress. At the peak, 80 percent of the laborers on the factory floor during the war were foreigners, compared with an average among German industries of 30 percent. Later the company would pay reparations to surviving slave laborers, while insisting that the blame lay with the Nazi regime and not the company. Porsche and Piëch family members also complained about the portrayal of Ferdinand Porsche by Mommsen, accusing the historian of trying to tar them with their grandfather's deeds.

The first major Allied attack on the Volkswagenwerk came on April 8, 1944. Bombers of the U.S. Eighth Air Force dropped 146 tons of explosive and incendiary bombs, which killed thirteen people, including four forced laborers. Areas used for offices suffered most of the damage. Two more raids followed in June, shortly after D-day, as part of Operation Crossbow, a joint British and American attempt to knock out sites that produced the V-1 rockets that Germany was using to terrorize Britain. Some of the forced laborers were shot to death by German guards for "plundering"—presumably trying to take advantage of the chaos to steal food. A fourth major raid, by eighty-five American

B-24 bombers, took place on August 5. This time most of the factory buildings suffered damage.

But though the bombs put holes in the brick walls of the factory and caused roofs to cave in, the effect on production was only temporary. Much of the machinery and production had already been moved to safety beginning in 1943, either to the basement of the factory itself or to other locations. For example, an iron mine in Tiercelet, in northeastern France, was converted into a factory to produce air-to-ground bombs that had previously been made at the Volkswagenwerk. The Volkswagenwerk continued to turn out the Jeep-like Kübelwagen until days before the American Army reached the outskirts of the town on April 10, 1945.

On that day, Jean Baudet, a laborer from France, noted in his diary that the road passing by the Volkswagenwerk was clogged with ragged, unarmed columns of German soldiers, many of them wounded. He could hear artillery and bursts of machine-gun fire in the distance. American fighter planes zoomed by at treetop level. But the fighting never reached the factory itself. The factory's force of armed guards melted away, and no one resisted when the first American tanks rolled onto the grounds. On April 12, Baudet looked out the window of the factory and saw a tank with a white star on it. "The Americans!" someone cried. Even the German civilians seemed relieved, Baudet wrote.

Ferdinand Porsche was long gone by then. In January, resigned and demoralized, he had retired to the family estate in Zell am See. His son-in-law, Anton Piëch, followed a few days before the Americans arrived, choosing not to fulfill his duties as commander of the local Volkssturm, the poorly trained citizens militia, armed with *Panzerfäuste*, that was supposed to mount a last-ditch defense of the fatherland.

Even after the Americans arrived, Pïech and Porsche continued to try to manage the factory from Austria. They saw themselves as custodians of Volkswagen, if not the outright owners. Piëch took a substantial amount of cash—about 10 million reichsmarks ($1 million)—from company coffers when he fled to Zell am See ahead of the Allied advance. In the months after Germany's surrender, Piëch sent bills to the British administrators of the plant for development work that Ferdinand and Ferry Porsche had done, and ostensibly continued to do, on the Volkswagen sedan and other projects, including an electric vehicle. Piëch and the Porsches even dismantled a barracks in Wolfsburg and transported the lumber to Zell am See, where the building was reassembled. When Volkswagen's postwar administrators tried to get Porsche to pay for the materials, Porsche refused, saying the lumber had been moved as part of a wartime plan to relocate Volkswagen headquarters. Porsche said it had cost the firm more money to re-erect the building than the materials were worth. These somewhat bizarre attempts to retain administrative control over Volkswagen came to an end when Ferdinand Porsche, Ferry Porsche, and Anton Piëch were arrested. The British did not pay the bills they submitted. But the behavior of the Porsche family members demonstrated their determination to retain a claim on the company that, at least in their view, Ferdinand Porsche had built.

The factory resumed operations even before the German surrender a month later, serving as a vehicle repair facility for the U.S. Army. Within a few weeks, it was also drawing on existing stores of parts to manufacture Kübelwagen again, this time for Allied use. By June, when British troops took over the factory because it lay in their zone of occupation, 133 vehicles had been produced.

The British quickly recognized that the Volkswagenwerk could help them solve two problems: the need to provide Germans in the area with a livelihood, and the British Army's own need for transport. The British prevented the machinery from being shipped away to one of the victorious countries as reparations, the fate that befell Opel, whose machinery was trucked to the Soviet Union. The red brick factory buildings were hastily repaired. Materials damaged by bombing were reused if they were still structurally sound. To this day, shrapnel scars are visible on some steel beams inside the plant.

The British renamed the young city Wolfsburg after a nearby castle, to remove associations with the past. In December 1945, production resumed, this time with the Volkswagen car as the main product. By

March 1946, one thousand had been produced and were deployed by occupation forces. As tensions with the Soviet Union increased, the British began to see the Volkswagenwerk as a way to provide well-paying jobs, to counteract the allure of communism, and to promote a democratic Germany. The Soviet zone, and the future border of East Germany, was less than ten miles to the east of the factory. British administrators, some with experience in the automobile industry, helped ensure supplies of precious raw materials and provided credit. "The responsible officers quickly realized that they could build not only autos at the factory but also democracy," the historian Markus Lupa has written. Very quickly the factory established a service and dealer network and by August 1947 had begun exporting vehicles to other countries in Europe. Volkswagen's international expansion had begun.

Like other German industrialists, Ferdinand Porsche was interned after the war. In August 1945, he was held for two months at a camp in Bad Nauheim, north of Frankfurt, known as Dustbin. Porsche's fellow prisoners included Albert Speer, the architect who was Hitler's chief of wartime production, and Wernher von Braun, who oversaw Hitler's rocket program and later played a crucial role in the U.S. space program. After being questioned by Allied interrogators about the Volkswagenwerk, Porsche was released on September 13, 1945, and returned to Austria. But his freedom was short-lived. Two months later Ferdinand Porsche, Ferry Porsche, and Anton Piëch were invited by the French government to Baden-Baden, a spa town in southwest Germany, to discuss the possibility that they would take over management of a French automobile factory. After being treated to a dinner in their honor, the three were unexpectedly arrested for their role in deporting French workers to the Volkswagenwerk during the war. Ferry was released in March 1946, but Ferdinand and Piëch were taken to Paris, where they were housed in the servants' quarters of a villa belonging to Louis Renault and put to work designing a new Renault passenger car. The French, fraught with internal divisions of their own, could not decide whether to exploit Porsche's knowledge or charge him with war crimes. In February 1947, as French officials pushing for prosecution gained the upper hand, Porsche and Piëch were put in a police jail south of Paris and later imprisoned in Dijon. However, Porsche benefited from testimony by Peugeot managers who said he had hindered deportation of the company's workers to Germany and protected the French carmaker from Nazi meddling. In August 1947, after twenty months of incarceration, Porsche and Piëch were freed. They were never tried for war crimes.

Ferdinand Porsche's influence over Volkswagen faded after the war, but he had already made two decisive contributions to its future development. He had designed the car that would be the basis of its postwar success. And he had passed on some of his passion for engineering and a sense of mission to a grandson, Ferdinand Piëch.

CHAPTER 3

Renaissance

IN THE FIRST FEW YEARS after American tanks clattered onto the factory grounds, Volkswagen's survival as a company was by no means a given. VW faced severe shortages of nearly everything needed to produce an automobile. Steel and other raw materials were rationed. The wartime dependence on forced laborers, who looted and vandalized the factory after liberation, then scattered, meant the company had an inadequate base of experienced workers. For the German workers who were available, there were not enough places to live. Many slept in the barracks once used by forced laborers. Food was so scarce that undernourished workers sometimes collapsed at the assembly line. Waves of German-speaking refugees, who had fled or been expelled from Soviet-controlled Eastern Europe, swarmed to Wolfsburg in search of work. But often they did not stay long, preferring to move to places where conditions were better.

Despite such travails, the Volkswagenwerk had a number of competitive advantages. Its factory had suffered less damage than those of competitors like Opel and Daimler. The factory had its own coal-fired power plant and was spared the frequent outages that handicapped manufacturers elsewhere in Europe. Because Volkswagen's inventory of machinery had largely escaped bombing and confiscation, the company could produce parts it could not buy from suppliers. The Volkswagen

Presswerk, the giant stamping machines used to form sheet metal into body parts and other components, was one of the largest in Europe.

Volkswagen was also fortunate that its British occupiers assigned a resourceful and energetic major named Ivan Hirst to oversee the factory. Hirst came from a family that had owned a British watchmaking company, Hirst Bros & Co., but had been forced to sell it in 1927 after sales collapsed, in part because of low-cost competition from Germany. Hirst had spent time as an exchange student in Berlin before the Nazis seized power. He admired the German work ethic and spoke a little German. For much of the war, he was attached to the Twenty-Second Advanced Base Workshop, a unit of the British Army's Royal Electrical and Mechanical Engineers Corps that operated just behind the front lines repairing damaged tanks and other motor vehicles. There Hirst acquired both crisis management skills and technical knowledge that would prove useful to Volkswagen.

A bookish-looking man who wore a dark, bushy mustache and round spectacles with black frames, Hirst arrived in Wolfsburg in August 1945. He effectively became Volkswagen's first postwar chief executive. The energy he brought to the task derived partly from Britain's interest in reviving Volkswagen and other German companies in order to reduce the cost of occupation. But Hirst, just twenty-nine at the time, also seems to have genuinely identified with the factory and its workers. Thanks to his talent for scrounging parts and raw materials, the company was able to produce its 10,000th postwar Volkswagen by October 1946, far more civilian vehicles than the Nazis ever managed. At Hirst's initiative, Volkswagen established a service and sales organization and made its first exports outside Germany—five cars delivered to a dealer in the Netherlands in October 1947. He pushed to improve the quality of the vehicles and to teach workers to take initiative rather than passively wait for orders, as they had been conditioned to do under the Nazis. Hirst also created a twelve-person workers council in line with the broader Allied goal of encouraging democratic institutions in occupied Germany. These were crucial steps in Volkswagen's evolution from a Nazi prestige project to an international, customer-oriented automaker. The decision to give workers a voice would set a precedent as well and have long-term consequences.

HIRST ALSO RECRUITED Volkswagen's first postwar German chief executive, Heinrich Nordhoff, who had managed one of Opel's main factories during the war. The urbane Nordhoff, who wore double-breasted suits with a white handkerchief poking from the breast pocket, had a deep understanding of how factories worked and was a capable manager. He led Volkswagen for two decades and oversaw a period of extraordinary growth. One of Wolfsburg's main thoroughfares is named after him. But Nordhoff later showed scant gratitude toward Hirst, without whom Volkswagen might not have survived as a corporate entity. Nordhoff insisted later that Volkswagen was nothing but a ruin when he arrived in November 1947 and had to be rebuilt from scratch. Perhaps that is how it looked to him. But Nordhoff, who took office on January 1, 1948, was lucky to arrive just as the West German economy was poised to take off. The introduction of the deutsche mark later that year, replacing the devalued reichsmark, gave Germans a currency they could trust, freed entrepreneurial spirits in the Western occupation zones, and dramatically improved the economic situation. Soon after the currency reform, production of Volkswagens more than doubled, to 2,500 a month.

Hirst left Volkswagen in 1948 but remained in Germany in other posts with the British government. In 1955, when Britain declared the occupation of Germany over and reduced its forces, Hirst found himself without a job. He returned to Wolfsburg, hoping there might be something for him to do. Nordhoff brushed him off. Only decades later did historians recognize Hirst's role in guiding Volkswagen through the perilous period immediately after German surrender. In the years before his death in 2000, Hirst became a frequent guest of honor at Volkswagen events in Wolfsburg.

By October 1949, when the British turned over administration of Volkswagen to the German government, Volkswagen had ten thousand employees. The four thousand cars a month they built represented half of the country's total vehicle production. Volkswagen was the main German supplier of vehicles to the British occupation forces and had a quasi-monopoly supplying the German postal service and the Deutsche Reichsbahn, the company that operated the German railroads. In 1949, 15 percent of the vehicles produced in Wolfsburg were sold abroad.

In many ways, Volkswagen and the German economic recovery that began after currency reform, the *Wirtschaftswunder*, were synonymous. As Germany tried to reclaim its place among the civilized nations of the world, the Volkswagen served as a mechanical ambassador of goodwill. Rather than conquering foreign lands by force, the Volkswagen charmed its way abroad. It was practical, reliable, and accessible to the burgeoning middle classes of Europe. Hitler's role in the creation of the car became a mere footnote, if it was remembered at all. Instead, Volkswagen emerged as a symbol of the new, democratic Germany. As German companies tried to rebuild the tarnished "made in Germany" brand, Volkswagen was in the vanguard. The factory's location just a short drive from the border with East Germany, and the "iron curtain" dividing Europe, only enhanced Volkswagen's symbolic value in the Cold War competition between free-market capitalism and Soviet communism.

In fact, the distinction between economic systems was not quite as great as it seemed. Volkswagen remained a state-owned company until 1960. Nor was Nordhoff a particularly committed democrat. During one meeting of top management in 1951, Nordhoff proposed that Volkswagen back candidates for local government to "make sure there are people in the city council who are willing to cooperate and won't create annoyance and difficulty." At another meeting, he suggested that the company insist that all newspapers and magazines that wrote about Volkswagen or its cars submit their articles in advance for examination and approval "so that unfriendly articles will no longer be published."

The KdF Wagen, now known as the Volkswagen, went on to be produced in numbers that far surpassed Ford's Model T and exceeded even the wildest expectations of its prewar planners. Conceived by a totalitarian government bent on military conquest, the people's car lived up to its name only when there was peace, when Germany was a democracy and Western ally, and the company had access to markets worldwide. Ferdinand Porsche died in 1951, at the age of seventy-five, long enough to see the Volkswagen become commonplace on German roads.

THE CONNECTION BETWEEN the Porsche family and Volkswagen endured after the war. Despite the accusations against him, Ferdinand Porsche and his family managed to keep a grip on the license for the Volkswagen design. Their legal claim, based on prewar contracts, was morally dubious. The Volkswagen had been ordered up by Hitler, financed by money confiscated from workers, and built initially in a factory staffed by forced laborers. The Porsche family saw it as a matter of protecting the patriarch's lifetime achievement from confiscation by the Allies. The Porsche engineering bureau also continued to do development work for Volkswagen under the direction of Ferry Porsche. From a business perspective, the Porsches emerged from the war well positioned to build a fortune that would make them one of Europe's wealthiest families.

Nordhoff set the precedent. In 1948, the Volkswagen chief executive negotiated a contract that gave the Porsche design bureau a 1 percent royalty on the first 500,000 deutsche marks of Volkswagen sales and 1 mark per vehicle after that. Administrators for the British occupation forces objected, questioning whether Ferdinand Porsche really owned the patents to the Volkswagen car. They also noted that a Jewish prewar shareholder of Porsche had filed a claim for restitution. That was a reference to Adolf Rosenberger, a businessman and race car driver who had provided start-up funds for Ferdinand Porsche's engineering office in 1931. But when the issue of patents ownership came before a court in Stuttgart, Volkswagen management argued in favor of the family. The millions that Anton Piëch had taken to Zell am See in the closing days of the war were credited to the bills that the Porsche design bureau had submitted. A letter from a Volkswagen executive named Hermann Knott to Ferry Porsche in 1949 illustrated the German management's sense of debt to Ferdinand Porsche: "The whole sphere in which we work, and the work we are doing, is in the final analysis a product of his spirit."

As soon as the British ceded control over German property in 1949, Volkswagen began paying money to the Porsche and Piëch families. The Porsche design bureau also received a contract and monthly retainer to continue performing development work for Volkswagen, which did not yet have a research and development department of its own. In addition, the family had exclusive rights to sell Volkswagen cars in Austria. The franchise may not have seemed terribly important at the time. But Louise Piëch, the determined and resourceful daughter of Ferdinand Porsche and Ferdinand Piëch's mother, took over management of the auto sales business after the death of her husband, Anton. She built it into the largest Volkswagen dealer network in Europe.

While members of the Porsche family defended their claim on Ferdinand Porsche's inventions, they fiercely rejected an attempt by Rosenberger, the Porsche financial supporter, to reclaim a share of the property he had been forced to give up under Nazi duress. In 1949, Rosenberger returned to Germany seeking compensation for his share in the design bureau as well as outstanding debts he said Porsche had still owed him when he fled Germany in 1935. Ferry Porsche refused. Rosenberger sued, but his chances of restitution dwindled as the Western Allies loosened their control over Germany. Left to pursue his claim in unsympathetic German courts, and suffering from declining health, Rosenberger, who had changed his name to Alan Arthur Robert after moving to the United States, settled in 1950 for 50,000 deutsche marks, about $12,000 at the time, and a free car. He died in Los Angeles in 1967.

Personal ties were as important as contractual agreements in the relationship between Porsche and Volkswagen. Ferry Porsche, who managed the family business after Ferdinand Sr. and Anton Piëch were detained by the Allies, took pains to stay on good terms with the postwar management of Volkswagen. In October 1953, after dinner at the Nordhoffs' in Wolfsburg, Ferry gushed with admiration that Heinrich Nordhoff had designed the family home himself without the help of an architect—a sign, Porsche wrote in a thank-you note, that Nordhoff had a fine sense of aesthetics. Porsche used the compliment to segue into a plea for Nordhoff to see a vehicle prototype that the design bureau had constructed. Nordhoff should come to Stuttgart and not only drive the prototype but run his hands across its perfectly formed sheet metal, Porsche wrote. In 1959, the affinity between Volkswagen and Porsche was further strengthened when Nordhoff's twenty-two-year-old daughter, Elisabeth, married Ernst Piëch, grandson of Ferdinand Porsche and older brother of Ferdinand Piëch.

Workers also retained a strong say over Volkswagen management, which was partly a legacy of the injustice the labor unions had suffered when the Nazis confiscated their funds and used the money to build the Volkswagenwerk. When the British gave up administrative control of Volkswagen in 1949, they passed ownership to the German federal government, which delegated control to the state of Lower Saxony. In 1960, the free-market oriented government decided to sell 40 percent of VW on the stock market. Workers, fearing the influence of profit-oriented investors, tried to block the sale. As a political compromise, Volkswagen employees won special concessions later enshrined by the German parliament. According to what became known as the Volkswagen law, decisions to open new factories or transfer work to different locations required a two-thirds vote of the supervisory board. Since workers had half the seats on the board, the law gave them veto power. In addition, the law restricted the voting rights of any one shareholder to a maximum of 20 percent, no matter how many shares the investor might hold. The law also guaranteed that the state of Lower Saxony and the federal government would each have two seats on the twenty-person supervisory board, as long as they owned at least one share.

The effect of all these provisions was to restrict the power of outside shareholders while ensuring that federal and local governments would be able to influence Volkswagen for as long as they chose. When the federal government later gave up its influence, Lower Saxony held tightly on to its seats on the board, which were usually occupied by the prime minister and the state economics minister. The state political leaders were typically from the left-of-center Social Democratic Party and almost always aligned with workers. The end effect was that Volkswagen workers from the 1950s through the 1980s probably had more real power over company policy than their comrades a few miles away in the so-called workers' paradise of East Germany. Outside shareholders at Volkswagen had essentially no say.

Meanwhile, Volkswagen technology served as the basis of the sports car company that Ferry Porsche built after the war, and which became the most important source of family wealth. The first Porsche sports car to be widely sold, the two-seat 356, was built atop a modified Beetle chassis and powered by a version of the Beetle's four-cylinder, air-cooled engine that had been upgraded to produce twice the amount of horsepower. The 356 and its best-known successor, the 911, which has gone through many iterations and is still being produced, were popular among sports car aficionados as well as jet setters. Porsche owed its popularity in America partly to a tragedy. In 1955 James Dean, the actor and symbol of youthful rebellion, died at the wheel of a Porsche 550 Spyder, a racing version of the 356, after the car collided with a Ford sedan that turned into his lane in Cholame, California. The Ford's driver may have overlooked the low-slung Porsche. Dean's death proved, morbidly enough, to be good for sales. By sports car standards, Porsches were also practical—less finicky than a British MG or Triumph and, because of their Volkswagen mechanics, relatively easy to maintain. Thanks largely to James Dean, Porsches were cool. Janis Joplin, the 1960s rock singer, drove a 356 convertible painted in psychedelic colors. The United States would remain Porsche's most important market for the next half century.

It would be almost four decades before a Porsche descendant again reigned in Wolfsburg. But there was never a period when the fates of the Porsches and the Wolfsburgers were not closely intertwined.

CHAPTER 4

The Scion

FERDINAND PIËCH, THE GRANDSON who had dreamed of working in the factory, was still a boy when the war ended. He was fifteen in 1952 when his father, Anton Piëch, died of a heart attack, at the age of fifty-eight. Ferdinand attributed his father's early death in part to the aftereffects of incarceration, in particular the abrupt transition from prison food to the rich table at the family estate. Otherwise, Piëch had much less to say in his autobiography about his father, who had managed the Volkswagenwerk, than about his more famous grandfather. A rebellious teenager, Ferdinand was sent by his widowed mother to a strict boarding school in the Alps. By his own account, Ferdinand had a reading disability that made learning foreign languages especially difficult. But he graduated and as a reward received a Porsche 356. Soon thereafter, he wrecked it while speeding along an Alpine pass. Like his grandfather, Piëch loved fast cars and pushing them to the limit.

In 1959, Piëch enrolled at the Eidgenössische Technische Hochschule Zürich, better known as the ETH. It was, and still is, one of the best technical universities in the world. Besides providing Piëch with a first-class education in engineering, the professors at the ETH also forced him to confront the reality of Nazi crimes. Teachers in Austria had avoided the subject. Piëch said later that he first heard the word "Auschwitz" during a required history course at the ETH.

Piëch does not seem to have been tortured by his family's association with Hitler, and he never condemned his father or grandfather. If he had any political convictions, he did not express them publicly. But he was at least willing to acknowledge Nazi atrocities.

During his years at the ETH, Piëch was already busy procreating, another activity that would characterize his life. He married his pregnant girlfriend, Corina, as he was beginning his studies. By the time he earned his degree in engineering in December 1962, at twenty-five, Piëch had three daughters. They turned out to be the first of many children.

Piëch hoped to work with airplanes after graduating from the ETH. But he was unable to get a job at one of the British or American companies that then dominated the aircraft industry. Porsche blamed his Austrian citizenship, but that seems doubtful, given the willingness of the Allies to forgive the dubious personal histories of the likes of Wernher von Braun, provided they had needed technical know-how. Even though Piëch never became an airplane designer, his studies of aeronautical engineering inspired a lifelong fascination with lightweight construction that he often applied to automobiles.

Piëch had already demonstrated a precocious aptitude for automotive mechanics. For his graduation project at the ETH, Piëch took the front and rear ends from two different prewar models made by Austro Daimler, a defunct manufacturer his grandfather had once managed. Piëch was able to fit the two halves together to make a single roadworthy vehicle. He was affronted when a professor suggested that he must have had help from Porsche engineers. They had better things to do than help the founder's nephew earn his diploma, he said.

Piëch does not seem to have had an urgent need to work at first. Despite being responsible for a young family, he was comfortable enough to spend the first three months after his graduation on the ski slopes. But he had an ambitious mother and had inherited his grandfather's drive. When the ski season ended in the spring of 1963, Piëch joined the fast-growing sports car company his uncle Ferry Porsche had founded in a section of Stuttgart known as Zuffenhausen. Thus began his career in the auto industry.

Piëch took charge of the Porsche racing program. At that time, Volkswagen supplied two-thirds of the Porsche racing budget, which served to enhance the image of both brands by financing cars to compete at tracks such as Le Mans or in the rally races that took place on European back roads. There was just one condition: all Porsche racers must have air-cooled engines. The requirement had little to do with technology and everything to do with marketing. By 1965, more than one million Beetles had been sold. Though there had been continual improvements, in essence the Volkswagen Beetle was the same KdF Wagen that Ferdinand Porsche had presented to Hitler almost three decades earlier. The design was beginning to seem anachronistic, in part because of the air-cooled motor. Among other problems, Volkswagen Beetles were notoriously frosty in winter, because they lacked the circulating fluid of a water-cooled engine that could serve as a medium to transfer heat from the motor to the car interior. But if Porsche cars were winning on the racetrack with air-cooled motors, the reasoning was, Volkswagen could still maintain that its technology was superior to the water-cooled motors used by almost every other manufacturer.

Win Porsche did, in part because Piëch showed he was willing to take great risks and spend grand sums. Piëch built his strategy around lightweight construction. The approach was particularly successful on the perilous mountain courses that then accounted for a large part of the racing calendar. Especially when driving uphill, the lighter the car, the faster it would go. To save a few pounds, some Porsche racers had aluminum fuel tanks that would easily split open in a crash, engulfing the car in flames. By Piëch's account, four Porsche racers died during his time as head of the racing program, but none because of his designs. Three died in cars by other manufacturers, and a fourth because of a poorly marked mountain course, according to Piëch. In those days, he said, the high death rate among drivers was regarded as an unavoidable trade-off. Still, Piëch was also showing the lengths he was willing to go to in order to win.

Within the Porsche-Piëch clan, however, Ferdinand Piëch's disregard for budget limits, at least when it came to building state-of-the-art automobiles, was stirring resentment. Eventually the resentment reached a crisis point, providing an early example of the family feuds that could roil the company.

By 1970, four of Ferdinand Porsche's grandchildren, out of a total of eight, held high positions at Porsche, the company. Porsche was by then a leader in the niche market for two-seat sports cars. It became known for vehicles that were not only fast and agile but also reliable and reasonably comfortable, and suitable for everyday use. The company had ended production of the 356 in 1966 after selling almost 78,000. It was replaced by the 911, which preserved some of the 356's characteristics such as an air-cooled engine in the rear and a hood that sloped between two protruding headlamps. But the 911 was more powerful and made concessions to safety such as the roll bar on the Targa convertible version. Some versions even had automatic transmissions, previously taboo for sports car purists. The 911 would be Porsche's core model for decades to come. Ferry Porsche was the chief executive. Ferdinand Piëch was the head of development. His cousin Ferdinand Alexander Porsche was head of design, and Peter Porsche, another cousin, was head of production. A fourth cousin, Ernst Piëch, was comanager of Porsche Austria, the sales organization. Though its products were world renowned, Porsche was still a relatively small company in 1964, with about 2,400 employees, who produced a little more than 10,000 vehicles.

In Ferdinand Piëch's telling, the family dispute began with a disagreement between him and Peter Porsche about motor technology and quickly escalated into an existential debate about the role of the family inside the sports car company. In the fall of 1970, in an attempt to resolve the crisis, Ferry Porsche assembled the entire clan at the family estate in Zell am See under the guidance of a group dynamics counselor. Instead of achieving harmony, the Porsches fought even more. The upshot was an agreement that everyone should quit the company, which would be turned over to professional managers. Ferry would still play an oversight role as chairman of the supervisory board. Ferdinand Piëch left the sports car maker the next year, after nine years at the company.

Back then, all that was at stake was a relatively small maker of sports cars. In the decades that followed, the family's wealth and influence would expand exponentially. But its ability to get along would not. Ferdinand Piëch was often at the center of family feuds, if not the instigator. Certainly, he did nothing to promote familial goodwill when, in 1972, he began an affair with Marlene Porsche, wife of his cousin Gerd Porsche. Ferdinand Piëch and Marlene lived together for twelve years, during which time he also had two children with another woman. At family dinners their table place names read "Frau P." and "Herr Piëch." These facts were not uncovered by some investigative reporter but are conveyed, with a sense of bemused detachment, by Piëch himself in his autobiography. He was sometimes startlingly candid about his own behavior, for which he offered no apologies.

So this was the future chief executive of Volkswagen at thirty-five. Piëch had shown himself to be a brilliant and daring engineer, perhaps too daring. He did not shy from a power struggle. In his personal relationships, he did not feel bound by societal norms.

Now barred from the family business, Piëch needed another job. Before long, he found one at Audi, a division of Volkswagen, formed in 1969 from the merger of two carmakers it had acquired, Auto Union in Ingolstadt and NSU Motorenwerke in Neckarsulm. Audi was then struggling to establish its own identity. Piëch was offered a job at the level of a department head, which he saw as a humiliating step down in status from his position as chief of development at Porsche, the position he held at the time he left. Piëch realized, he wrote later, that he had to prove to outsiders that his success at Porsche was not due solely to family connections. Audi was based in Ingolstadt, a city on the Danube between Nuremberg and Munich, about 325 miles south of Wolfsburg by car. It was far from the beating heart of Volkswagen. The parent company then regarded Audi as little more than a production facility. It is likely that no one at Audi appreciated the full significance of the new employee. A grandson of Ferdinand Porsche, sharing many of the founder's talents and personality traits, and perhaps some of his flaws, was again in their midst.

Years later, Ferdinand Piëch ridiculed attempts to psychoanalyze him in the reflection of his famous grandfather, whom he described as a distant figure who was usually absent working on his numerous projects. But the two had much in common, including fearless determination, a fascination with technology, and a genius for engineering. Both defined success in terms of the technical innovations they achieved. Both enjoyed the good life, but neither showed much interest in money for its own sake. Wealth was a pleasant byproduct, not an end in itself. Both were accused of paying insufficient heed to the consequences of their behavior on other people. When all was said and done, the grandson rivaled and perhaps topped the grandfather in terms of his influence on the German automotive industry, and on the company called Volkswagen.

CHAPTER 5

Chief Executive

WHILE FERDINAND PIËCH WAS PREOCCUPIED with the Porsche racing program and finding a place for himself in the automotive industry, Volkswagen was becoming one of the world's preeminent carmakers and a symbol of Germany's rebirth.

During the 1950s, Volkswagen began exporting its cars outside Europe to countries such as Brazil, Australia, and South Africa. Simultaneously, the company built up the dealer and service networks needed to sustain foreign markets. The United States received a first shipment of 330 vehicles in 1950. Six years later, Volkswagen sold 43,000 Beetles in the United States and another 6,700 Volkswagen vans, which had been introduced in Germany in 1950. As sales in the United States grew, Volkswagen considered producing Beetles at a former Studebaker factory in New Jersey. But it concluded that the plant would not be profitable. In those days, wages in the United States were too high compared with the cheap labor available in Germany.

For Volkswagen and many other German companies, exports were essential to survival. American carmakers did not need to export, because their home market was so big. But German companies quickly bumped up against the limited purchasing power of their domestic market. To stay competitive with their American rivals in an industry where size brought considerable cost advantages, German carmakers had to find markets abroad—or die. The international reach of German manufacturers remains the basis for the country's economy today. Volkswagen helped show the way.

Volkswagen also played a part in building the mythos of Germany as a land of meticulous craftsmen and engineers. An early ad by the New York agency Doyle Dane Bernbach told of Kurt Kroner, an inspector in Wolfsburg, who refused to let a 1961 Beetle leave the factory because a chrome strip on the glove compartment was blemished. "This preoccupation with detail means that the VW lasts longer and requires less maintenance, by and large, than other cars," the copy said. The underlying message was that, whatever you might think of the Germans, you could trust them to build a car. In decades to come, Volkswagen managers cultivated this image of obsessive attention to detail, elevating it practically to a cult. They would discover that it could also be a liability.

Improbably, considering that the Volkswagen had originally been ordered up by the Nazis, the Beetle became a symbol of the counterculture. This was due in no small part to the efforts of Doyle Dane Bernbach, which Volkswagen hired in 1959. The company had done no advertising in the United States before then. Carl Hahn, the Volkswagen executive in charge of the U.S. market, went shopping for an agency after dealers complained that the company was not doing enough marketing. Hahn toured Madison Avenue, where he was wooed lavishly by teams of account executives and copywriters at the big established agencies. When he got to Doyle Dane Bernbach, the agency partner Bill Bernbach met him alone in a windowless room. As Hahn later remembered the meeting, Bernbach showed him a couple of the agency's campaigns, including one for El Al Airlines, and sketched out a vision of how to position Volkswagen as an icon of simplicity and reliability. Hahn liked Bernbach's lack of pretension and hired him.

In the land of Buick and Oldsmobile road yachts, the four-cylinder Volkswagen had to find another way to appeal to the emotions of car buyers. The purchase of an automobile is never a purely rational decision. With DDB's "Think Small" campaign, Volkswagen made compactness a virtue, something to brag about. The self-deprecating tone of the ads marked a complete departure from the Madison Avenue practice of bludgeoning consumers, and helped revolutionize the advertising business. "The rounded fenders were, in effect, the biggest tail fins of all," Bob Garfield later wrote in *Advertising Age*. "For what Volkswagen sold with its seductive, disarming candor was nothing more lofty than conspicuously inconspicuous consumption. Beetle ownership allowed you to show off that you didn't need to show off."

The pitch worked. In 1964, Volkswagen exported 330,000 cars to America—sales figures it would strain to replicate decades later. The company opened a factory in the north German port of Emden, where Beetles could be loaded onto cargo ships bound for America as soon as they rolled off the assembly line. Volkswagen even had its own specially built cargo ship, the *Johann Schulte*, which could carry 1,750 vehicles to overseas markets. During the 1960s, Volkswagen was the best-selling foreign-made auto in the United States. As protests against the Vietnam War gathered force, Volkswagen tapped into the backlash against consumerism and the military industrial complex, with which the Detroit automakers were identified. (Robert McNamara had been president of Ford before becoming secretary of defense during the Vietnam era.) Volkswagen Beetles and vans, often painted with peace signs and psychedelic colors, were ubiquitous among the lines of cars backed up on the way to the Woodstock festival in 1969. Unlike any other car, the Beetle had personality. It is hard to imagine any other vehicle capable of taking on anthropomorphic qualities and starring in a movie, as the Beetle did in the 1969 Walt Disney caper *The Love Bug*. The transformation of the Volkswagen from Nazi propaganda project to counterculture phenomenon was one of the most spectacular examples of rebranding in the history of marketing.

Germans were probably less sentimental about the Beetle, known in German as the *Käfer*. In the United States, the Beetle addressed a certain market segment. In Germany, the Beetle defined middle-class transportation, accounting for one-third of all newly registered passenger cars in 1964. The members of the seminal electronic music band Kraftwerk depicted a Beetle (as well as a Mercedes-Benz sedan) on the cover of its 1974 album, *Autobahn*. But the droning, repetitive lyrics—"Wir fahr'n fahr'n fahr'n auf der Autobahn . . ." (We're driving, driving, driving on the autobahn . . .)—sounded more like a comment on industrialized sameness and the soulless nature of consumerism. Decades later, when Volkswagen tried to tap nostalgia for the Beetle with the New Beetle, which evoked the contours of the original Ferdinand Porsche design but featured modern mechanics, the car sold well in the United States but poorly in Germany.

On February 17, 1972, a few months before Ferdinand Piëch began work at Audi, Volkswagen passed a historic milestone. Production of the Beetle reached 15,007,034 cars, overtaking Ford's Model T as the most-produced car ever. It was perhaps the ultimate vindication of Ferdinand Porsche's original vision. But the vision was running out of road. In 1972, Volkswagen sales declined by 14 percent, to 1.5 million vehicles. Of those, 1.2 million were Beetles, an illustration of how dependent the company was on a single model line.

Volkswagen sales in the United States peaked at 570,000 in 1970, about a third of the total production. In 1972, they fell to 486,000. The Beetle was finally displaying signs of mortality, and there was not yet anything to replace it. By then, thirty-four years had passed since a Nazi advertising campaign had offered Germans the opportunity to pay for a Kraft durch Freude Wagen in up-front installments. (In 1961, the West German government offered 600 deutsche marks toward a Beetle or 100 marks in cash to people who had made payments on a KdF Wagen and never received a car.) Although the Beetle had been continually refined, with larger motors and amenities such as automatic transmissions, the fundamental design was still the same. Doyle Dane Bernbach did its best to present the Beetle's quirks as virtues. A print ad in 1970 showed the spacecraft that landed on the moon with the headline "It's ugly, but it gets you there." There was no other copy, just a Volkswagen logo. Other ads claimed that Volkswagen left the body design the same so that it could spend the money on mechanical improvements.

In fact, Volkswagen had grown complacent. Heinrich Nordhoff, who remained chief executive until 1967 and died in 1968, had initially not thought much of the Beetle but then was slow to push for something to follow it. After the introduction of a van in 1950, which won its own place in the 1960s counterculture, there was no completely new model until 1961, when Volkswagen began selling the 1500, a more conventional-looking sedan. The 411, a midsize car that came in hatchback and station wagon versions, went on sale in 1968. Both the 1500 and the 411 had air-cooled motors mounted over the rear wheels, an increasingly outdated technology despite Volkswagen's attempts to use the Porsche racing program to portray it otherwise. Neither of the new cars was a huge success. As a result, Volkswagen was vulnerable when the Beetle's popularity inevitably began to sag.

Much of the design work on prospective Beetle replacements was done by Porsche. Besides producing sports cars, Porsche engineers continued to function as Volkswagen's de facto research and development department. Even though members of the Porsche and Piëch families did not own substantial quantities of Volkswagen stock or hold positions in top management, the company remained the wellspring of their expanding wealth, as well as an important part of their identity. Likewise, generations of Volkswagen managers recognized a de facto moral claim on the company by the descendants of Ferdinand Porsche, despite his close association with Hitler.

Legally, the sports car company in the Zuffenhausen neighborhood of Stuttgart continued to be known as Dr. Ing. h.c. F. Porsche AG. The name stood for Doctor of Engineering, honorary, Ferdinand Porsche Stock Corporation. (The family was eager to preserve the patriarch's academic distinctions, even if they were honorary. Ferdinand's formal education was limited to his training as a pipe fitter.) The company occupied the same buildings and was the same corporate entity as the engineering and design bureau that Ferdinand Porsche had operated before the war, and where he had designed the Beetle. Its main shareholders were Louise Piëch and Ferry Porsche. Porsche's attempts to design a successor to the Beetle in the 1960s and 1970s never amounted to much. (Ironically, when Volkswagen finally launched the Golf in 1974, the body was designed not by Porsche but by an Italian, Giorgetto

Giugiaro.) But there were other successful collaborations. One was the 914, a quirky two-seater aimed at first-time sports car buyers that went on sale in 1970. The 914 was, in fact, the first Porsche-designed Volkswagen to come to market since the Beetle. The other prototypes for passenger cars were rejected.

The 914 was sold as a VW-Porsche in Germany and simply as a Porsche in the United States. The car's motor was mounted just behind the passenger compartment, roughly midway between the front and rear axles. The mid-engine configuration improved weight distribution and handling. The 914 was fast and fun, with a removable plastic "targa" top for summer driving, and it was affordable. A basic version could be had for 12,000 deutsche marks, or well below $4,000 at the time. In 1975, the 914 was replaced by the 924, which had a water-cooled motor mounted under the front hood and a hatchback in the rear that, by sports car standards, provided at least some room for luggage. Because Porsche's production capacity in Stuttgart was limited, the 924 was built by Audi at its factory in Neckarsulm, less than forty miles to the north. In fact, the car was arguably more Audi than Porsche. But if buyers were aware of the distinction, they did not seem to care. The 924 was a success. By the time production stopped in 1988, more than 150,000 had been produced.

The cooperation between Volkswagen and Porsche taught both companies an important lesson in marketing. Their partnership demonstrated that it was possible for models to share many components and yet retain their distinct brand identities. This was true even if it was an open secret that a car sold with a Porsche badge was made at an Audi factory. What mattered to buyers was that the car looked and drove like a Porsche. This principle would later become an important part of Volkswagen's growth, as well as a profitable business model for Porsche.

Over at Audi, Ferdinand Piëch was busy proving that he was not just Ferdinand Porsche's grandson but also a formidable engineer and manager in his own right. In the early 1970s, Audi did not yet have the upscale image it acquired later and was struggling to establish its own identity within the Volkswagen corporation. There was considerable tension between the parent company in Wolfsburg and the subsidiary in Ingolstadt. Piëch's attitude toward Wolfsburg was dismissive, and his autobiography gives no indication that he identified with the parent company or was concerned about how Volkswagen would resolve its dependence on the Beetle. Piëch saw the parent company as authoritarian—an accusation later thrown at him—hierarchical, and prone to quashing innovation.

One of Piëch's first assignments at Audi, as it happened, was to solve a problem in the United States. The Audi 100, the company's top-of-the-line sedan, was unable to pass U.S. emissions tests, which meant the new model could not be sold there. Indeed, Piëch noted in his autobiography, the car was technologically incapable of meeting U.S. standards. It was an early example of the problems that Volkswagen had in reconciling U.S. and European emissions rules. The U.S. Clean Air Act of 1970 required automakers to reduce emissions of carbon monoxide and hydrocarbons, such as cancer-causing benzene, by 90 percent by the 1975 model year. Nitrogen oxides had to be reduced by 90 percent by the 1976 model year. The European Union, then known as the European Economic Community (EEC), did not have unified air quality standards until 1980. European pollution limits were generally not as strict as those in the United States, and the fragmented nature of the EEC meant that enforcement was uneven.

Traveling to the United States, Piëch solved the emissions problem by negotiating a deal with government regulators. They would give the Audi 100 temporary approval in return for a promise to equip the engines with fuel injectors within six months. The improved efficiency of the injectors, which sprayed fuel directly into the cylinders rather than mixing it with air inside a carburetor, would bring the necessary reduction in emissions.

Piëch's reward for solving the American problem was to be promoted to head of testing at Audi in early 1973, a position that made him a candidate for head of development, should that position ever open up. Unexpectedly, it did later that year when Piëch's boss, Ludwig Kraus, quit rather than bow to a dictum from Wolfsburg that he report to the Volkswagen head of technology, Ernst Fiala.

To his chagrin, Piëch didn't get the job right away. He was passed over for a rival, Franz Behles. Piëch also thought about quitting, but instead made a deal with top management at Volkswagen. Piëch would stay, and if Behles didn't work out after a year, Piëch would become development chief. That is exactly what happened. Behles was blamed for his department's failure to develop a practical version of the Wankel rotary motor. The Wankel, with its rotating combustion chambers, was simpler and more compact than a conventional piston-driven engine and offered excellent acceleration. But its fuel economy was substantially worse. As Piëch well knew, Behles faced a mission impossible. To this day, no one has been able to develop a rotary engine economical or practical enough to win a substantial share of the mass market. But at Volkswagen the price of failure was dismissal or demotion. Behles was sidelined. Piëch, who proposed a much more efficient five-cylinder motor as an alternative, became Audi's head of development.

It was an early example of Piëch's brash approach to corporate politics. With the freedom that comes from being independently wealthy, Piëch could afford to take career chances that people with mortgages might not dare. He was confident of his own engineering ability and not afraid to pit his skills against those of rivals. In his autobiography, Piëch rarely mentions allies or mentors. He often writes about rivals or people who stood in his way. He conveys the impression of someone who struggled to the top by using his engineering ability and infighting skills, rather than by winning friends and building networks.

As head of development at Audi, Piëch was later remembered for his role in bringing four-wheel drive to passenger sedans. Audi in the 1970s was groping for ways to set itself apart from the mass-market products of its parent company and compete for the same well-heeled customers as BMW and Mercedes-Benz. At the time, four-wheel drive was seen as the province of off-road vehicles. The extra gears and drive shafts needed to power a second pair of wheels added weight, which meant the motor had to be more powerful to deliver the same performance as a two-wheel drive car. There didn't seem to be any point in putting four-wheel drive in a car intended solely for the road.

According to Piëch, the idea that four-wheel drive might improve handling and traction in passenger cars came from Jörg Bensinger, Audi's head of chassis testing. Bensinger had noticed the superior performance on icy and snowy roads of a prototype four-wheel drive vehicle being developed for the Bundeswehr, the West German army. Characteristically, Piëch did the design and development work for the first four-wheel drive Audi sedans in secret, to avoid any meddling from Wolfsburg. Piëch revealed a prototype at a meeting of top Volkswagen managers in the mountain village of Turrach, Austria. Without announcing beforehand that the car had four-wheel drive, Piëch and team drove an Audi 80 up a snow-covered ski trail in view of the gathered dignitaries, who were duly astonished.

Piëch named the technology "quattro" with a small *q*. The cars came on the market in 1980. That and other technical innovations, such as use of aluminum body parts to reduce weight, or coating the undercarriage in zinc to protect against rust, helped establish Audi as a technological pioneer and allowed it over time to surpass BMW and Mercedes-Benz in European sales. The evolution of Audi's image from producer of staid and bourgeois middle-class cars to maker of sporty and innovative luxury vehicles was crucial to the parent company's future. The rarefied premium segment yielded profit margins simply not possible in the overcrowded mass market occupied by Volkswagen brand cars. Over time, Audi would produce far more than its share of the parent company's profit, helping to protect Volkswagen from the losses that plagued rivals like Fiat, General Motors' Opel unit, or Ford of Europe. Much of the credit belongs to Piëch.

By then Piëch was in his forties. He had gone bald, but was still fit-looking and handsome despite protruding ears. He was rich and he was powerful. By 1984, he had nine children: five with his first wife, Corina; two with Marlene Porsche; and two more from a woman Piëch declined in his autobiography to identify. The relationship with Marlene ended after she hired a twenty-five-year-old governess for the children. Piëch first tested the job candidate's ability to handle a four-wheel drive vehicle on the steep road to the family's mountain chalet. She passed. The new governess was a former kindergarten teacher named Ursula Plasser. Soon after she started to work, she and Piëch began an affair. They married in 1984. Ursula Piëch, with whom Ferdinand Piëch would have another three children, bringing the total to twelve, was a vivacious foil for her dour husband, and was game for the pastimes he enjoyed, like sailing in the Mediterranean and skiing, though she was less excited about his penchant for fast Japanese-made motorcycles. Ursula Piëch, better known by her nickname Uschi, embraced the automobile-centric world of her husband and would go on to play a prominent role not only in her husband's life but also at Volkswagen.

Meanwhile, at Audi, Piëch was working on a technological innovation that would have far-reaching consequences for Volkswagen. It was one of the initiatives of which Piëch was later most proud: diesel engines designed for passenger cars. Invented in the late 1800s by Rudolf Diesel, a German, diesel had long been in widespread use in trucks and ships, where it provided superior fuel economy and longer engine life. But it was much more difficult to deploy diesel in smaller vehicles.

In all automobile engines, much of the potential energy is wasted because the fuel does not burn completely. In a diesel engine, the fuel—distilled from petroleum in a process different from that for gasoline—is compressed with oxygen inside the cylinders until it becomes so hot from the pressure that it ignites. In a gasoline engine, by contrast, spark plugs cause the fuel to ignite. Diesel engines are less wasteful because the dense, highly compressed fuel and air mixture burns more thoroughly than in a gasoline engine. As a result, diesel engines go farther on a gallon of fuel than a gasoline motor. The disadvantage of diesels, at least until Volkswagen and other automobile makers began civilizing them in the 1970s, was that they tended to be louder and smellier than gasoline engines, and they vibrated more, making the cars less comfortable to drive. In addition, the diesel combustion process produces more torque and places more strain on the moving parts inside an engine. The components of a diesel engine have to be strong enough to take the punishment, which means they are also heavier. All of these problems needed to be solved in order to make diesel practical for passenger cars.

The impetus for the development of diesels for passenger cars came from the oil crisis that began in 1973. Arab oil producers imposed an embargo in retaliation for U.S. support of Israel. The price of oil quadrupled, the average price of gasoline rose 40 percent, and voluntary rationing led to long lines at filling stations. Carmakers were forced to pay more attention to fuel economy. Diesel potentially offered a way to improve fuel efficiency without a penalty in performance, if the technology could be tamed enough for use in cars. In 1976, Volkswagen offered a diesel version of the Golf, the water-cooled hatchback introduced two years earlier as the company struggled to replace the Beetle. But though the diesel Golf won fans, it was too noisy and shaky for most buyers, and certainly not refined enough for the affluent customers Audi was trying to attract.

Typically, Piëch began work on a diesel motor for Audi behind the backs of his superiors at Volkswagen. If they knew about it, they would get in the way. But the project was discovered by a Volkswagen executive during a visit to Audi headquarters in Ingolstadt and Piëch—as he had feared—was forced to cede control to Wolfsburg. Undeterred, he continued to pursue development of diesel for Audi separately. Of course, it made no sense from a cost point of view for two sets of engineers inside the same company to be working separately toward the same goal. Piëch admitted that he forbade such duplication of effort when he later moved to Wolfsburg. When he was still in Ingolstadt, though, diesel was too important to him to relinquish control.

Even in those days, when regulations were much less strict, Piëch worried about how to get a handle on diesel emissions. That was another of the downsides of diesel engines. Diesel emitted several nasty pollutants, in particular nitrogen oxides that promoted the formation of smog and could cause or exacerbate health problems. Diesel also produced fine soot particles that could penetrate deep into human lungs and cause cancer. But it became possible to reduce, though not eliminate, these alarming drawbacks by deploying a combination of technical innovations. Chief among them were advanced injection systems that could more precisely control the timing and volume of fuel being delivered to the cylinders. Meanwhile, a turbocharger improved delivery of air to the cylinder. Combined with an electronic motor control system, which drew on advances in computer technology, the turbocharger and fuel injector could tailor the combustible mixture of air and fuel to whatever demands were being placed on the engine. Whether the car was idling, climbing a hill, or speeding on a highway, the fuel-air cocktail could be mixed and burned for the best result. The improved motor still had a bit of a growl, but it was no longer prone to blowing clouds of black exhaust like the diesels of old.

Volkswagen called the result TDI, or turbocharged direct injection. It took eleven years to perfect. Audi unveiled its first TDI model, an Audi 100 sedan, at the Frankfurt Motor Show in September 1989, a few weeks before the fall of the Berlin Wall. Piëch was proud of the innovation, which used an onboard computer to manage the engine, then a novelty. The five-cylinder engine in the Audi 100 used two liters less fuel per 100 kilometers of driving than the competition, he bragged. At the same time, the car accelerated more quickly and ran more cleanly. Emissions were 30 percent less, according to Piëch. (He did not define which emissions he was referring to, however.) In his view, TDI had revolutionized the image of diesel in the world, and Audi enjoyed a big head start in bringing it to market. As the man who had driven the development of TDI, Piëch could take much of the credit. For him, it was a professional milestone, one that gave him a personal stake in diesel technology. Piëch would continue to be a forceful advocate of diesel for the rest of his career.

Volkswagen was not the only passenger car maker to produce turbocharged diesel engines. Mercedes, BMW, Fiat, and Peugeot, among others, used similar concepts, and in some cases were years ahead of Audi and Volkswagen. Mercedes-Benz sold a turbocharged diesel sedan beginning in the late 1970s. But from a technological point of view, TDI was a particularly elegant combination of fuel injection, turbocharging, and electronics that could be produced cheaply enough for midrange cars. The smart performance and excellent fuel economy won fans among drivers. In the United States, there was even a TDI Club of Volkswagen diesel enthusiasts. Perhaps more than any other automaker, Volkswagen succeeded in turning its diesel technology into a widely recognized brand name. Diesel Volkswagens were sold with a metal TDI logo stamped on the rear of the vehicle.

Certainly, TDI helped Piëch make a name for himself within Volkswagen and positioned him to become chief executive of the Audi division. He got the job on January 1, 1988, after once again displaying his skill at corporate intrigue and his willingness to engage in brinksmanship. Angry at plans to extend the contract of the person who held the job before him, Wolfgang Habbel, Piëch threatened to resign. He considered himself the better man for the job and did not want to wait. By his own account, Piëch had tense relations with his fellow managers and with Carl Hahn, who had become chief executive of Volkswagen in 1982. But Piëch wagered that Volkswagen would not dare allow someone of his technical prowess to jump to a competitor, and he was right. The Audi supervisory board revoked Habbel's contract extension and gave Piëch the top job instead.

To prevail, Piëch drew on support from the worker representatives, who would prove to be an important source of his power in years to come. In line with the German law known as *Mitbestimmung*, or codetermination, worker representatives held half the seats on the Audi supervisory board. (The chairman, who represented shareholders, had a tie-breaking vote.) Because of its history as an independent company, Audi still had its own supervisory board, which formally chose the members of the management board, though of course Wolfsburg exerted strong influence. Piëch communicated his threat to resign not only to Hahn but also to Fritz Böhm, head of the Audi workers council. The council, separate from the union but closely associated with it, represented employee interests within Audi. Böhm, who had worked at Audi since 1950, was especially powerful. He was a socialist who had fought in street brawls against the Nazis before Hitler took power. As a soldier during the war, Böhm was sent to the eastern front, captured by the Red Army, and forced to work in a Soviet car factory. He was a tough character. The workers council led by Böhm, who was also on the Audi supervisory board, had de facto veto power over management decisions. Piëch, with his fine sense of power relationships, learned early to get the workers on his side. (When Böhm celebrated his ninetieth birthday, in 2014, Piëch was among those who spoke at the ceremony honoring him.) With Böhm's support, Piëch took the helm of Audi.

The promotion made Piëch a prime candidate to become chief executive of Volkswagen when Hahn's contract was scheduled to expire five years later. Piëch's advancement was regarded with ambivalence by his colleagues. Hahn commented that Piëch's ascension to the top job at Volkswagen could no longer be hindered—not exactly a ringing endorsement of his ambitious subordinate. By Piëch's own estimation, he did not fear conflict and had no compunctions about firing or demoting people who in his view were not performing. In 1985, when a brake disc failed on the Audi car competing in a rally race on the Mediterranean island of Corsica, Piëch summarily removed the manager in charge of the team. The manager found another job inside the company, Piëch noted drily. (In Germany, labor regulations make it hard to fire people outright. It is often easier to send someone to the corporate equivalent of Siberia.)

When Piëch took over as chief executive of Audi, one of his first moves was to push the chief financial officer into retirement and assume control over the purse strings himself. Piëch saw the man as overly obsequious toward Wolfsburg. Piëch's ambition was to outshine the mother company. Despite quattro, TDI, and other innovations, there was a lot of work to do. Audi was not yet the cash cow it would later become. Its profit margin was a meager 2 percent, it had too many workers, and it had fallen behind the Japanese carmakers in manufacturing efficiency. Piëch immediately assembled a meeting of the top layers of management and put the executives on notice that underperformance would not be tolerated. And he managed to force through job cuts, which was difficult but not impossible if conditions were dire enough. To survive, Audi needed to bring its costs more in line with those of competitors like Toyota, which set the standard in manufacturing efficiency. In all, four thousand people lost their jobs. Piëch had no trouble firing bad managers, but he later said that he could never forget the weeping family members of Audi workers who gathered outside his door. It was a rare expression of sentimentality by Piëch, and therefore its sincerity is open to question. But it is also true that Piëch avoided mass layoffs during the rest of his career. He justified his unforgiving approach toward the managers who worked under him in terms of job preservation. Better to fire an incompetent executive than let the person endanger the jobs of innocent assembly line workers.

All of the problems that Audi had were present at the mother company, and then some. They became acute sooner than expected. By the early 1990s, Volkswagen had long since switched to cars with water-cooled motors mounted in the front of the car, like virtually every other mass-producer of automobiles. Since the decline of the Beetle, Volkswagen had struggled to build a car that was as original or as popular. The first water-cooled Volkswagen passenger car to achieve lasting success in the market was the Passat sedan, introduced in 1973. Soon thereafter, Volkswagen launched the sportier Scirocco and the compact Golf, which was intended to serve the same market for economical, practical cars as the Beetle. The factory in Wolfsburg produced its last Beetle on July 1, 1974. (Versions of the original, air-cooled Beetle designed by Ferdinand Porsche would continue to be produced at other sites for several more decades. The last—the 21,529,464th—was made in Puebla, Mexico, on July 30, 2003.) The new Volkswagen models sold well in Europe, and the Golf eventually exceeded the Beetle in total sales. But Volkswagen continued to lose ground in the United States, outclassed by Toyota, Nissan, Honda, and other Asian brands in the market for fuel-efficient, reliable, affordable cars. Even in Europe, growth was halting. Volkswagen vehicle sales crossed the 2 million mark for the first time in 1969. It wasn't until 1990 that sales reached 3 million vehicles.

Volkswagen production spiked to 3.5 million vehicles in 1992, but the increase was illusory. In the euphoria that followed German reunification, Volkswagen produced far more cars than it could sell. When reality set in, and it became clear that East Germany and countries in Eastern Europe would take decades to catch up with the West, Volkswagen found itself with acres of unsold inventory. Hahn was a salesman who had established Volkswagen in China and taken advantage of the opening up of Eastern Europe to acquire the Czech automaker Skoda. Both moves would later bring huge payoffs. But under Hahn, Volkswagen had not kept pace with advances in productivity pioneered by the Japanese. Because of high labor costs and inefficient factories, the company made practically no money on the cars it did sell. The United States was a disaster for Volkswagen. Amid a worldwide recession, its sales in the United States fell to a little over 100,000 vehicles, the lowest in decades. Among the shortcomings that Piëch blamed on Hahn, one was significant in light of later events. Volkswagen lacked early warning systems to ensure that management became aware of serious problems like these before it was too late to fix them. The same would later be said of Volkswagen under Piëch. (Hahn believed that he had been made a scapegoat for Volkswagen's efficiency gap, which he blamed on the workers' council and the IG Metall union and their power to block job cuts.)

In 1992, Volkswagen's overall profit fell to just above the break-even point, and it was clear to insiders that the next year would bring big losses. There was even talk that Volkswagen would go bankrupt.

Piëch's hour had come. He had demonstrated his toughness and willingness to cut costs at Audi. Moreover, he was an engineer, with an obsession for automobiles and manufacturing that had been a part of his being since he was a young boy riding the train inside the Volkswagenwerk past rows of Luftwaffe bombers undergoing repairs. Volkswagen's unexciting model lineup needed the technical excellence that Piëch had brought to Audi. Even workers recognized that Piëch, despite his history as a job cutter at Audi, was the inevitable choice.

Because of Volkswagen's unusual shareholder structure, no person could become chief executive without the support of the workers. Volkswagen had been listed on the stock exchange since 1960. But the state of Lower Saxony, where Wolfsburg is located, retained a large block of shares as well as two seats on the supervisory board. The politicians almost always voted with labor, in part because the state was usually controlled by the center-left Social Democrats, and also because it was difficult for a politician of any stripe to vote against the perceived interests of workers at the state's largest private employer. Workers already had half the seats on the supervisory board, in line with German labor law. Together with the representatives from the government of Lower Saxony, they easily commanded a majority.

There were attempts by other members of the supervisory board to recruit alternatives to Piëch from outside Volkswagen. But in the end the choice came down to Piëch and Daniel Goeudevert, a Frenchman who was Volkswagen's head of sales. The representatives of labor and Lower Saxony met in Kassel, a city about a hundred miles southwest of Wolfsburg. Both candidates made their pitches. By Piëch's account, Franz Steinkühler, chairman of IG Metall, the powerful steel and auto industry union that represented Volkswagen workers, called him at about 9 p.m. at his hotel room in Kassel to say his chances did not look good. After all, Piëch had battled with workers over Audi layoffs. Piëch accepted the bad news with customary coolness and decided he would look for work elsewhere. Then he went to sleep. About 2 a.m., Piëch was awoken by Steinkühler. Piëch was labor's unanimous choice. It was the equivalent of the head of the United Auto Workers choosing the president of General Motors. The decision by the supervisory board in Piëch's favor several weeks later was little more than a formality.

Piëch took office on January 1, 1993. He was well aware that he was regarded with apprehension, if not hostility, by many people in the company. "Only when a company is in severe difficulty does it let in someone like me," he wrote later. "In normal, calm times I never would have gotten a chance."

CHAPTER 6

By All Means Necessary

THERE WAS LITTLE DOUBT that Volkswagen was badly in need of change when Piëch took up residence in Wolfsburg on January 1, 1993. His office was on the top floor of the thirteen-story brick high-rise that overlooked the vast factory complex and was topped with a Volkswagen logo that looked like a giant hood ornament. Carl Hahn, Piëch's overly optimistic predecessor, had allowed the workforce to balloon to 274,000 people in anticipation of growth that never came, including almost 120,000 employees in Germany, far too many for the firm to operate profitably.

In the United States, the Japanese had all but driven Volkswagen from the market with cars that were more economical and more reliable. By the time Piëch took office, Volkswagen had slipped to a humiliating fifteenth place among importers in the United States, the world's largest auto market and one no carmaker could ignore. In addition, Volkswagen was paying substantially more for parts than competitors because contracts were often awarded on the basis of traditional relationships rather than competitive bidding. The system for delivering components from suppliers to the assembly line had become dysfunctional.

Toyota and other Japanese rivals had pioneered just-in-time supply chains, in which parts were delivered to the factory just before they were needed. This reduced the need for warehouse space. It also cut the amount of time that elapsed between when the company paid for a part and when it sold the vehicle. As a result, less money was tied up in the production process. At Volkswagen, by contrast, parking lots near the factory were clogged with unfinished vehicles awaiting some crucial component, wasting both real estate and capital.

The scope of the problems became obvious in 1993, Piëch's first year in office. Amid a worldwide economic downturn, sales plunged 10 percent, to 76.6 billion deutsche marks ($46.8 billion), and the company reported a 1.9 billion mark ($1.2 billion) loss. There was talk of insolvency. It was a crisis but, for an ambitious manager like Piëch, also an opportunity. Any clever executive trying to make a name for himself would rather take over a struggling business than a healthy one. The potential for glory is greater at a company in need of turnaround, assuming the situation is not hopeless, and at Volkswagen it wasn't. If things are already running smoothly, there is less room for improvement, but plenty of opportunity to be blamed if business turns sour.

As an engineer, Piëch was far better qualified to address Volkswagen's efficiency deficit than his sales-oriented predecessor. Long an admirer of the Japanese, Piëch set about copying their manufacturing methods in much the same way that his grandfather had copied Ford when planning the Volkswagenwerk in the 1930s.

New factories in Mosel, in East Germany, and in Martorell, a Spanish city near Barcelona, presented blank slates where Piëch could try out Japanese-style production methods. Under Piëch, Volkswagen copied the Toyota *kaizen* system of assembly line teams, where all workers were encouraged to continually look for ways to improve quality or save time and money. For example, female workers in a Volkswagen upholstery department complained that it was difficult to pull the covers over headrests. After employees and supervisors met to discuss the problem, someone found an unused machine that was modified to compress the foam padding on the headrest, so the cover could be pulled on faster and with less effort. Even if such improvements were tiny, they could add up to huge gains in efficiency over time. The new production techniques were phased in as Volkswagen introduced new models or new generations of existing models, a gradual process because it usually takes four to seven years to design a new car and bring it to production. The company held workshops at its factories where workers, dressed in identical light gray overalls, learned the new manufacturing philosophy.

A similar transformation was taking place at Porsche's main factory in Stuttgart. As is so often the case with makers of desirable, expensive sports cars, Porsche was not always profitable. Sports cars are luxury goods rather than necessities, and sales can plunge steeply during economic downturns or stock market crashes when potential buyers decide to cancel or postpone purchases. That was especially true before the fall of the Berlin Wall in 1989. Until the 1990s, when markets in Eastern Europe, Russia, and China began to open up, Porsche was dependent on Europe and the United States. When both suffered recessions in the early 1990s, Porsche sales plunged so precipitously that the company suffered three money-losing years in a row and was close to bankruptcy.

Like Volkswagen, Porsche recognized that Japanese auto manufacturers were much more efficient and that, to survive, it needed to copy their methods. Under Wendelin Wiedeking, a brash production expert whom the Porsche and Piëch families recruited as chief executive in 1992, Porsche imported former Toyota managers and engineers to teach the secrets of *kaizen* manufacturing at the factory in Stuttgart. Under Wiedeking, Porsche expanded its product line to include the Boxster, a mid-engine, two-seat convertible that won raves for its handling and helped reduce the company's dependence on the 911. Starting at under $50,000, the Boxster was about $30,000 less expensive than the basic 911 and within reach of a greater range of the dentists, accountants, and other professionals who made up Porsche's customer base. The manufacturing efficiency drive and Boxster, as well as the elimination of 1,850 jobs, or about 20 percent of the workforce, returned Porsche to profit by 1996.

Another pillar of Piëch's efficiency drive was the intensification of sharing among different vehicles. The management buzz phrase was "platform strategy." Collections of parts, everything from engines to seat adjustment mechanisms, were used among as many vehicles as possible. Often the sharing was invisible to customers. The electric motor that adjusted the side mirror on a Golf, for example, was the same one installed in a top-of-the-line Audi and all the models in between. Few customers knew or cared.

The savings from parts sharing were enormous. Instead of having to design a dozen different mirror adjustment motors, it was necessary to design only one. Instead of ordering a dozen motors from suppliers, negotiating on price, drawing up contracts, and so on, it was necessary to order only once. (In practice, Volkswagen often bought parts from several suppliers in order to encourage competition among them.) The platform strategy maximized what are known as economies of scale. Generally speaking, the per-unit cost to make something goes down as the overall volume increases. A million mirror adjustment motors cost less, per motor, than 100,000. It's the automotive equivalent of saving money by buying a case of soda rather than just a single can. Managing the flow of parts from suppliers to assembly lines is also easier and cheaper, because there are fewer kinds of parts to keep track of.

The platform strategy offered huge advantages, but it also came with a big risk. If there was a problem with a component, the problem could spread like a virus through the whole company. Any vehicle that used the defective component would be infected. If, say, a particular engine was installed in eleven million cars, and that engine turned out to have a serious defect, the consequences for Volkswagen's reputation and finances were potentially enormous.

Piëch did not invent the platform strategy, but he forced its use at Volkswagen. Before the end of the 1990s, for example, Volkswagen was using a single platform for the Golf, the Audi A3 compact wagon, the Czech-made Skoda Octavia sedan, and the Audi TT sports coupe. It was the same principle, writ large, that Ferry Porsche had employed when, after the war, he used the chassis and motor of a utilitarian KdF Wagen as the basis for a sexy sports car, the Porsche 356. The approach came full circle in 1997, when Volkswagen began producing the New Beetle on the same platform as the Golf. Though the New Beetle was mechanically conventional, with a water-cooled motor mounted in the front, it referenced the body shape of the original KdF Wagen in an attempt to tap Beetle nostalgia.

Piëch also set about enlivening Volkswagen's solid but rather dull product lineup. This was perhaps the most important task of all, and one for which Piëch was well suited. In the auto industry, it is axiomatic that no amount of cost cutting, slick marketing, or financial legerdemain can compensate for boring products. Piëch began a drive to infuse the mass-market Volkswagens with some of the same technical showmanship he had brought to Audi. The goal was to give Volkswagens something that would set them apart in a European market crowded with brands selling to middle-class buyers, including Fiat, Renault, Peugeot, Citroën, Ford of Europe, and—Volkswagen's main German rival—the Opel unit of General Motors.

Soon the boxy Golf took on sleeker dimensions made possible in part by the use of a laser to weld the roof. Under Piëch's demanding eye, the workmanship improved. Piëch had a fetish—his word—for measuring the gap between pieces of sheet metal on the car, for example the space between the door and the body. Piëch's obsession with so-called body gaps had a logic to it. Narrow gaps gave the car a cleaner appearance and an aura of fine craftsmanship. Moreover, reducing the body gaps by a few millimeters forced the entire organization to focus more intently on quality. In order for the front hood to close snugly against the sheet metal covering the fenders, everything else had to fit perfectly—the hinges connecting the hood to the body, the front grill, the radiator, and so on throughout the whole vehicle. There was no longer room for error. To achieve the necessary precision, everyone in the organization had to coordinate more effectively, from the people who designed the car, to suppliers, to the supervisors on the assembly line. The focus on body gaps was, on the one hand, a bit maniacal and typically Piëch. But it was also a clever way to bring discipline to the manufacturing process.

Under Piëch, the idea of the people's car took on new meaning. Now it meant luxury car features and performance in products accessible to the masses. The GTI version of the Golf, for example, offered sporty performance and handling for around $19,000; auto magazines dubbed it the poor man's BMW.

True to character, Piëch pushed technical boundaries. Mindful of how oil crises had devastated the auto industry in the past, Volkswagen introduced a new subcompact, the Lupo, which went on sale in 1998. (It was never exported to the United States.) The TDI diesel version could travel 100 kilometers on three liters of fuel, or 78 miles per gallon, the first mass production car to be able to do so. (In Europe, fuel economy is measured in terms of how much fuel it takes to travel 100 kilometers.)

It did not take long for the manufacturing efficiency campaign to pay off. By 1997, Piëch's fifth year in office, the time required to build a Passat had fallen from thirty-one hours to twenty-two. Profit also recovered. Volkswagen passed the break-even point in 1994, and in 1997 profit once again topped 1 billion deutsche marks ($650 million). Sales in the United States bounced back, too, rising 55 percent in 1998, thanks largely to the New Beetle. Sales for Volkswagen, including Audi, reached 382,000 in Canada, the United States, and Mexico, the best result since 1974.

Remaking the assembly lines and products played to Piëch's strengths. But he also had to deal with Volkswagen's German workforce, a task that required a degree of political savvy and finesse—not his obvious strong point. Even Carl Hahn, much more of a diplomat and compromiser, had struggled to control organized labor's tendency to view Volkswagen as a full-employment scheme and to demand higher pay while pushing for shorter working hours. By the late 1980s, a typical Volkswagen worker earned monthly base pay of about 4,100 deutsche marks, or $2,630, for a thirty-six-hour workweek. Even after instituting a hiring freeze and pushing some workers into early retirement, the company had thirty thousand more people than it needed, according to the company's own estimates. Reducing the workforce would be especially difficult at Volkswagen. Though Piëch had become one of the most powerful auto executives in the world, he still served at the pleasure of the Volkswagen supervisory board, controlled by a coalition of labor leaders and the state of Lower Saxony. They were unlikely to sanction mass layoffs of the size that had taken place at Audi, in the state of Bavaria.

Piëch's solution to this quandary was innovative and, it must be said, humane. He recruited Peter Hartz, a steel industry executive, as head of personnel. The son of an ironworker, Hartz had been a member of the IG Metall auto and metal workers union earlier in his career and was a Social Democrat. Square-jawed and stern, with neatly trimmed gray hair and rimless oval eyeglasses, Hartz negotiated an agreement with the company's German employees to work a four-day week. The work week was cut to 28.8 hours instead of 36. There was a corresponding cut in pay, to about 3,280 deutsche marks per month ($2,025), and employee representatives and unions agreed to be more flexible about working different shifts or jobs, which was crucial to Piëch's plan to remake manufacturing along Japanese lines. The plan cut costs without putting workers out on the street.

That innovation earned Hartz a reputation well beyond Wolfsburg. The 1990s were a period of great national insecurity about whether Germany could remain competitive in a world that was much freer than it had been a few years earlier. The fall of the Berlin Wall had opened up trade with Eastern Europe and the former Soviet Union, which presented an opportunity for German exporters. But the emergence of China, with its low-cost labor, also raised questions about whether Germany and other wealthy countries could compete. Hartz had shown that there were ways to ease the human costs of globalization.

Perhaps there was an element of compassion in Piëch's use of the four-day week, spurred by memories of the weeping family members of laid-off workers he had seen at Audi. But the solution also reflected political reality.

At Volkswagen, few things are more sacred than workers' right to be at the table when major decisions are made. The partnership between management and workers is viewed not as a handicap but as an essential element in Volkswagen's rise and, indeed, an example to the world. The official company history gives almost as much weight to changes at the top of the *Betriebsrat*, the workers council, as to new chief executives. Managers and worker representatives might argue about a course of action, but once they achieved consensus everyone marched forward under the same proud banner—so the argument went. What other company could boast such a level of management-worker cohesion?

By law and tradition, German workers at all companies enjoy far more influence than their counterparts in the United States. A cooperative relationship between workers and management was a central tenet of Rhineland capitalism, the postwar economic model that was designed to be a humanistic riposte to East German–style socialism. A German law first passed in 1976 guarantees the rights of all employees at larger corporations to have a say in working conditions. The system is known as *Mitbestimmung*, or codetermination. The law gives employees the right to elect workers councils, which must be consulted before a factory can, for example, introduce weekend shifts or make job cuts. At any company organized as an *Aktiengesellschaft*, or stock corporation, workers hold half the seats on the supervisory board, which oversees top management. The chairman of the supervisory board, who represents shareholders, casts the tie-breaking vote if there is a deadlock, giving shareholders a de facto one-vote majority—except at Volkswagen.

Volkswagen employees had even more power because the state of Lower Saxony and its two representatives on the supervisory board—usually the state prime minister and the minister of economics—almost always voted with labor. They ensured that workers rather than shareholders maintained control over decisions such as whom to appoint as chief executive. No job cuts were possible without worker assent.

One of the few people in Wolfsburg who dared criticize the system was Carl Hahn. He blamed codetermination laws for destroying German competitiveness, and for German corporations' declining standing in industries they once dominated, like pharmaceuticals and electronics. "Often codetermination leads more than ever to destruction of jobs, and seldom to progress for those the union headquarters claim to protect and promote," Hahn wrote in his memoirs. As events would show, he was one of the few people who saw the system's potential for corruption.

Ferdinand Piëch did not criticize labor's power. He turned it to his advantage. Piëch and the workers were, in some ways, natural allies. Workers were most concerned with preserving their own jobs and pay, and creating jobs for their children. Piëch's goal was to push the boundaries of automotive engineering and turn Volkswagen into a global superpower. That ambition required lots of workers.

All modern manufacturing companies face continual tension between technology gains and employment. As automation and other efficiency improvements reduce the need for human labor, the only way for a company to avoid layoffs is to keep growing. The need to keep getting bigger was especially urgent at Volkswagen, because the company already lagged rivals like Toyota in worker productivity. To keep their jobs, Volkswagen workers required an empire builder, and they had one in Piëch.

Perhaps Piëch even felt a genuine kinship with workers. Like him, many of their fathers and grandfathers had also worked for Volkswagen. In any case, allying with workers allowed Piëch to run the company pretty much as he pleased. Unlike shareholders, the workers council displayed little interest in the details of company strategy. Its members were happy to let Piëch call the shots as long as their jobs were secure. "Neither side observed *Mitbestimmung* as a cooperative process in questions of company policy," wrote Werner Widuckel, a former high-ranking official in the VW workers council. "*Mitbestimmung* was primarily the policy of redistribution."

The shareholder representatives on Volkswagen's supervisory board were ambivalent about Piëch. He came to power because workers backed him, and he could not expect to rule Volkswagen without their continued support. For the rest of Piëch's career, workers would be a crucial source of his power. He could be dismissive toward fund managers and other outside shareholders, he disdained the financial press, and he quarreled with other members of the Porsche family. But he avoided alienating workers. The long-term risk from Piëch's strategy of worker appeasement was that he would fall into the same trap as his predecessors and allow Volkswagen to become bloated and overstaffed.

Piëch's aversion to mass layoffs did not extend to the brick tower in Wolfsburg that housed the executive offices. Unlike Hahn, who avoided making enemies, Piëch had no compunction about pushing out people he considered incompetent. Before the end of 1994, less than two years after he had taken office, Piëch had replaced almost the entire board of management, nine top executives in total. It is perhaps not surprising that Piëch would want to replace Hahn's team with his own. But his abrupt methods were a shock after the Hahn years. When Hahn took over as chief executive in 1982, one of his priorities was to loosen what he considered the overly militaristic management style established by Heinz Nordhoff after the war. Hahn pushed for more open debate in the management board, whose members were accustomed to quietly working out deals on major decisions, so that meetings were largely a formality. Under Piëch, the pendulum began to swing back in the other direction.

The fired managers included Piëch's own handpicked successor at Audi, Franz-Josef Kortüm, who was pushed out after only thirteen months in the job. Kortüm was blamed for heavy losses at Audi, which seemed a stretch. In the car business, with its slow-moving development cycles, it was hard to see how Kortüm could have ruined Piëch's former fiefdom in so short a time. In any event, Piëch had kept Audi on a short leash after moving from Ingolstadt to Wolfsburg, and he certainly shared the blame for any missteps. The German press took note of the new atmosphere. "A climate of fear and mistrust reigns in the management offices of Audi," *Der Spiegel* wrote. "Hardly a manager dares offer factual criticism of a Piëch decision."

David J. Herman, the American chief executive of Opel, General Motors' European unit, also noticed a change in mood. Once a month, it was the custom for the chief executives of German auto manufacturers to meet to discuss issues of common interest. Although the carmakers competed with each other, the atmosphere at the meetings was friendly. The other participants included Bernd Pischetsrieder, the jocular, Havana cigar–smoking chief of BMW, and Helmut Werner, the extroverted chief of Mercedes-Benz. In contrast to the convivial Hahn, Herman recalled later, Piëch was cool and standoffish. According to Herman, he would spend the meetings quietly taking notes in triple-size letters, writing in blue ink with a fountain pen, using a large notebook he kept in an aluminum briefcase. (The large letters were apparently a function of Piëch's reading disability.)

Piëch's chilly demeanor was a portent. In those days, Volkswagen and Opel were the main competitors for middle-class car buyers in Germany, the largest auto market in Europe, and the key to domination of the Continent. BMW and Mercedes-Benz targeted a more upscale market, but Opel and Volkswagen fought for the same customers not only in Germany but across Europe as well. Volkswagen would eventually overtake GM worldwide, but in 1992 the American carmaker was more than twice as large, producing more than 7.7 million vehicles a year, compared with 3.5 million for Volkswagen. In Germany, Volkswagen was market leader, selling 372,000 vehicles in 1992, compared with 289,000 for Opel. Piëch soon signaled that he wanted more than just market leadership. He wanted total dominance, and he was prepared to deploy new tactics to achieve it, whatever the consequences.

In late 1992, even before he formally became chief executive, Piëch began secretly courting José Ignacio López de Arriortúa, General Motor's star purchasing manager. López, who came from the Basque region of Spain, was famous in the industry for his ability to bargain down supplier prices and for his command of the intricacies of supplier networks. He was among the first people to recognize that suppliers could help advance automotive technology, and he rewarded those who were innovative. Volkswagen was desperate to lower the cost of its components, and López was credited with helping to rescue GM from financial disaster in the early 1990s. His management techniques were sometimes eccentric, which only enhanced his aura of genius. López and his aides wore their wristwatches on their right wrists, vowing not to wear them on their left wrists until GM was profitable again.

He insisted that his aides lived on a "warrior's diet" of fruit and whole grains, with only a little meat and no alcohol except for an occasional glass of wine. Even Piëch, a famously demanding boss, was awed by López's ability to get people to work to the limits of their endurance. Piëch and López met secretly in November 1992 in a room at a Sheraton hotel adjacent to the Frankfurt airport. Piëch was pleased to discover that López had studied engineering. They got along well.

Piëch courted López for several months, even promising to consider building a Volkswagen factory in López's beloved Basque homeland, something GM was no longer willing to do. (By some accounts, Piëch made a definite promise to build the factory. Piëch himself was vague about what commitments he made to López.) When rumors began to circulate in the press that López was considering defecting to Volkswagen, López denied them. In fact, he was secretly working out contract details with Volkswagen. When López on March 16, 1993, announced his decision to jump to Volkswagen, GM was enraged.

Poaching a rival's best executives is nothing new. But López's departure, along with seven members of his GM team, soon took on the dimensions of a scandal. GM accused López of carting away twenty boxes full of confidential documents when he left. According to a lawsuit that GM later filed against Volkswagen as well as Piëch, López, and other executives in federal court in Michigan, the documents contained vital trade secrets, such as lists of GM agreements with parts suppliers worldwide, prices, contract terms, and delivery schedules. GM claimed that the documents also included plans for an advanced factory and GM's model strategy for the next decade. Among other things, the documents would allow Volkswagen to see what GM was paying for parts and demand equal or better prices. According to the complaint, López and his team had spent a month copying the documents into Volkswagen computers, and then had shredded them, which GM charged was evidence of a systematic conspiracy. The suit further alleged that Piëch had induced López to steal the documents and accused Volkswagen of witness tampering because Volkswagen had allegedly suppressed incriminating portions of a report on the incident prepared at the company's request by KPMG Peat Marwick. Piëch denied GM's claims, saying he had never encouraged López to take documents. In any case, Piëch said, the boxes contained little more than books, magazines, brochures, seminar materials, and the like.

GM went so far as to claim that, under the Racketeer Influenced and Corrupt Organizations Act, or RICO, Volkswagen was a criminal organization. When Volkswagen lawyers argued that the RICO claim was outlandish and tried to get it dismissed, a federal judge refused and allowed the case to go forward. The evidence of industrial espionage was solid enough that German prosecutors and the U.S. Justice Department began criminal investigations.

Piëch did nothing to defuse the situation when he spoke of "Krieg" (war) on GM, and said that Volkswagen would defend itself "mit allen Mitteln" (with all means). The Opel chief Herman found it strange that Piëch did not qualify the phrase. He did not say "with all ethical means" or "with all legal means," just "all means." At a press conference on July 28, 1993, Piëch peered over a bank of microphones, a thin smile on his face, and expanded on the armed-combat metaphor. "Whenever there is a war there are fewer people around at the end," he said icily, his head swiveling slowly back and forth as he surveyed the news reporters. "There are always winners and losers. With our partners, that Volkswagen has around the world . . ." Here he paused. "I plan to be the victor."

Piëch later insisted that the word "war" had first been used by a *New York Times* reporter whom he was only quoting. But at least to non-German eyes, his performance at the press conference fed unfortunate stereotypes of cold-blooded Germans determined to win, whatever the body count.

Herman saw the press conference on German television. It confirmed his view that Volkswagen under Piëch was no longer content to continue the relatively cordial competition with Opel that had prevailed since the end of the war. He believed Piëch wanted to eliminate Opel from the market. "VW set out to destroy Opel," Herman said.

The court battle between GM and Volkswagen dragged on for years.

There were moments of absurdity. GM had trouble serving Piëch with the legal papers because he avoided traveling to the United States. According to Herman, the company learned that Piëch would be at a mountain lodge in British Columbia. GM dispatched a man to the lodge disguised as a backpacker, who managed to get on an elevator with Piëch and hand him the papers. GM sought as much as $4 billion in damages from Volkswagen, the *New York Times* reported at the time. But GM began to lose its appetite for pursuing the suit after polls showed it was hurting Opel's image among German buyers, many of whom sympathized with Volkswagen. In January 1997, the two companies agreed on a settlement. It called for Volkswagen to pay GM $100 million, and also agree to buy $1 billion in parts from Delphi, a GM subsidiary. (Volkswagen was already a Delphi customer.) Volkswagen did not admit wrongdoing but acknowledged "the possibility that illegal activities" might have been committed by some of the employees who had defected from GM. López agreed to resign from Volkswagen. The settlement was one of the largest ever in a corporate espionage case. Despite that, it was largely a face-saving deal for both companies exhausted by the endless legal proceedings.

In Germany, prosecutors in the city of Darmstadt, near Opel's headquarters in Rüsselsheim, pursued criminal charges against López. So did authorities in the United States. Neither of the investigations resulted in any grave penalties. German prosecutors settled with López in 1998. He was required to donate 400,000 deutsche marks, or about $190,000, to charity, and the charges were dropped. The German prosecutors concluded, according to a trade publication, *Automotive News Europe*, that the case was "too cumbersome, too complicated and was no longer in the public interest to pursue."

In the United States, a grand jury in Detroit handed down a six-count indictment of López in May 2000, accusing him of wire fraud and the interstate transportation of stolen property. But by then López was back in Spain, which refused to extradite him. The Spanish high court ruled in June 2001 that the charges—some of which carried maximum sentences of ten years in prison—were not serious enough to warrant extradition. The court also accepted López's argument that he was no longer mentally fit to stand trial. In 1998, López had been in a car crash near Bilbao, Spain, that nearly killed him and left him in a coma for three months. Appearing before the Spanish court in 2001, López, then sixty, had gained 130 pounds since he had last been seen in public, according to Emma Daly, the *New York Times* reporter who attended the hearing in Madrid. Though López bantered with judges during the proceedings, doctors testified that he suffered memory lapses and personality changes and had turned over control of his financial affairs to his wife. A psychologist hired by the U.S. Justice Department said that López was capable of testifying, but the Spanish court elected to let him stay in Spain.

And so the López affair came to an end. López never held an executive position at a major auto company again, though he continued to work as a consultant and pursue his dream—never realized—of opening an auto plant in the Basque region. Years later, business schools still used the affair as a case study in business ethics. For Volkswagen, the case should have at least served as a warning of the cost of overly aggressive business methods. It might have been an occasion to examine what controls existed to protect the company from wrongdoing by its managers. It could have given top management an opportunity to issue clear guidelines about what ethical limits Volkswagen employees should observe as they pursued the company's ambitions.

But Volkswagen did nothing in spite of a growing need to establish corporate standards. The 1990s was a period of proliferating government regulation of the auto industry in safety as well as in emissions. In the United States, the Clean Air Act was overhauled and significantly strengthened in 1990. The EEC introduced the so-called Euro 1 standard for autos in 1992. It set limits on nitrogen oxides and other tailpipe emissions in a more systematic way than in the past. The increase in official scrutiny sent a signal to many in the industry to bolster their internal compliance systems. But, as Volkswagen executives would later admit, the company lagged the rest of the industry in establishing such checks and controls.

In his autobiography, Piëch showed no sign of contrition about his conduct in the López affair. He could not, he wrote, perceive anything that Volkswagen or López had done wrong. Whereas López was forced to leave Volkswagen as part of the settlement with GM, one of his lieutenants, Francisco Javier Garcia Sanz, prospered. Garcia Sanz, had followed López from GM to Wolfsburg in March 1993, and joined the Volkswagen management board in 2001 as head of procurement.

Piëch emerged from the López scandal with his power undiminished, if not enhanced. The people who mattered most—the politicians from Lower Saxony, the worker representatives, and the shareholders on the supervisory board—were willing to accept Piëch's rough edges as long as he continued to deliver growth. If there was a loser in the whole controversy, it was Opel. Once Volkswagen's peer, and with a much longer tradition in Germany, the GM division bled money, and its market share in Europe began to dwindle. In 1992, Volkswagen, including Audi, Skoda, and SEAT (a Spanish carmaker acquired by VW in 1986), had a 17.5 percent share of the car market in Western Europe, according to the European Automobile Manufacturers Association. General Motors, including Opel and Saab (which belonged to GM), had a 12.5 percent share. In 2000, Volkswagen had an 18.7 percent share compared with 10.8 percent for GM. The hemorrhaging continued, until by 2013 Volkswagen brands had a 24.6 percent share in Europe and GM just 7.1 percent.

Whether López had much to do with Opel's deterioration is debatable. Certainly GM made many mistakes in Europe. The turnover among Opel managers was too high, and the executives GM sent from Detroit often lacked the engineering know-how necessary to compete with the formidable Germans. GM paid too little attention to the specific needs of European drivers, especially demanding German customers. There was chronic discord with Opel worker representatives. One of Opel's biggest lapses was its failure to respond to growing demand for diesel engines in passenger cars. In part because of lobbying by Volkswagen and other German manufacturers, diesel was effectively subsidized in Germany and other European countries. The sales tax imposed on diesel was lower, so it was always cheaper than gasoline at the pump, a powerful sales argument. Caught unprepared by the surge in demand for diesel engines, Opel had to buy motors from Isuzu. And Opel was up against one of the best automotive engineers of all time, a pioneer in bringing diesel to passenger cars: Ferdinand Piëch. "To say that he was man of the century in terms of engineering and style and the creation of the greatest automotive empire, would be a great understatement," Herman said.

But Herman also believed that Piëch's behavior established a new, more ruthless approach to competition. The signal that Piëch sent through his conduct in the López affair could not have been overlooked by the managers, salespeople, and engineers who worked for him. "If you want to trace the origins of the company ethic," Herman said many years later, "I think you need to go back to that period." Market domination was the one and only objective, by all means necessary.

CHAPTER 7

Enforcers

IN THE EARLY 1990S, about the same time that Volkswagen under Ferdinand Piëch was conquering Europe with its fuel-saving diesels, a few engineers in the United States and Europe were preoccupied with a related but inverse problem: how to make sure the increasing numbers of diesel vehicles on the road were clean. In 1990, sweeping revisions to the Clean Air Act passed both houses of Congress with large majorities and support from Republicans as well as Democrats. The legislation was an ambitious effort to radically improve air quality in the United States, particularly in urban areas where cars and trucks were the main source of pollution. But while it was easy for Congress to say that cars should produce fewer harmful emissions, it was not so easy to determine whether automakers were complying.

What comes out of the tailpipe of an automobile can vary radically, depending on many factors: the speed of the vehicle, the outside air temperature, whether the engine is warm or cold, whether the air conditioner is on or not, whether the driver is cautious or lead footed. The challenge for regulators was to measure emissions in a way that was consistent and would produce fair comparisons among the dozens of makes and models on the market.

The standard approach was to test cars in the lab on rollers. That made sense in a lot of ways. A great deal of equipment was required to capture engine exhaust and measure its chemical components. It was obviously easier to bring cars into a lab than to put a lab inside an automobile. A laboratory also made it possible to reproduce precisely the same conditions for every vehicle being examined. Cars standing on rollers could be subjected to the same simulated driving pattern, using a trained driver, without variables of terrain or weather. The risk of all those finely tuned parameters, though, was that a lab test would be artificial. It would do a fine job at showing how much a car polluted under controlled conditions, but say little about the vehicle's performance on the road. The other risk was that the testing procedure was by definition predictable. The speeds used to test cars, the temperature inside the laboratory, even the humidity—it was all public knowledge. It was a test for which automakers already knew the answers.

One of the people who had doubts about the testing methodology was a young engineer in the EPA's compliance division named Leo Breton. "Seeing a number of tests over a few years," Breton said, "I would scratch my head, not having been there forever, really wondering if this had anything to do at all with what vehicles were emitting in the real world."

Breton, who had a master's degree in mechanical engineering from the University of Maryland, began working at the EPA offices in Washington in 1991. His job was to oversee the private contractors who performed much of the testing. Washington was not the center of the EPA's automotive compliance universe. Most of the expensive testing equipment was in Ann Arbor, Michigan, close to Detroit and the headquarters of the American automakers. In Breton's view, the people in Ann Arbor were overly cordial toward the auto companies they were supposed to regulate. They had no interest in destroying the industry that their jobs depended on, he thought. But it was clear that auto companies needed watching. Complying with clean air rules is expensive. Automakers had a clear incentive to cheat, and a long history of doing so.

Breton was involved in uncovering one early case. In 1993, EPA testers discovered that cars made by General Motors' Cadillac division emitted almost three times as much carbon monoxide with the air-conditioning or heating systems on than when climate control was off. It turned out that, following customer complaints that '91 Cadillacs tended to stall, GM had installed a computer chip in the cars that increased the ratio of gasoline to air inside the cylinders when the climate control system was on. The chip fixed the stalling problem, but worked only by overriding the emissions control system. The result was higher output of carbon monoxide.

In legal terms, the computer chip qualified as a defeat device—a mechanism designed to reduce the effectiveness of the emissions system when the official testers weren't looking. As GM knew, the EPA disengaged climate control systems during emissions tests. By programming the emissions override to kick in only when the car's heat or air-conditioning was on, GM could be confident that the stalling fix, with its side effect of higher carbon monoxide emissions, would never show up in an EPA lab.

Or so GM thought. In fact, the EPA had been questioning its practice of not testing cars with air conditioners running. In 1993, the agency decided to do a study of how much changing the procedure would affect the results. The officials happened to pick a Cadillac to conduct experiments, and quickly noticed that something was odd. With the climate control system on, the cars emitted up to 10 grams of carbon monoxide per mile, compared with a legal limit of 3.4 grams per mile. A Justice Department investigation later concluded that 470,000 Cadillacs from 1991 to 1995, including Seville and Deville models, had defeat devices. The cars had released an additional 100,000 tons of carbon monoxide into the atmosphere, the Justice Department calculated. Carbon monoxide is extremely poisonous. It can cause heart and circulation problems, lead to headaches and impaired vision, and interfere with people's ability to do their jobs or learn. In high concentrations it is fatal. "These so-called defeat devices are not just paper violations," Janet Reno, the U.S. attorney general, said in November 1995, "but result in real increases in emissions that affect real people." GM agreed to pay $45 million to cover penalties, the cost of recalling and fixing the cars, and measures to offset the emissions, such as buying school buses that polluted less than older models. At the time, the settlement was the largest ever for violations of the Clean Air Act by a vehicle manufacturer.

The Cadillac case underlined the problems of testing cars only in labs. It also illustrated how the increasing computerization of engine controls opened up ways for automakers to cheat. Concerned about the potential for sophisticated deception, Breton started wondering: what if you could put the laboratory inside the car and see what cars were really polluting on the road? That was, after all, what mattered to the environment and the human beings who would inhale the emissions. Such a test would be less predictable and harder to defeat.

Within the EPA, though, there was no mandate for such a project. The prevailing wisdom within the industry at the time was that lab tests were accurate, and even if they weren't, they provided a level playing field to compare different vehicles and measure automakers' progress toward meeting progressively stricter clean air rules. The EPA already had an expensive testing center in Ann Arbor, and no interest in supporting research that might undercut the credibility of its investment. "The real world could only rain on the party," Breton said. According to Breton, the EPA didn't object when he said he wanted to develop a way to test cars on the road, but it didn't give him a budget either. There was no money for science projects, he was told.

Breton admits he was regarded within the agency as a bit of a loose cannon, though he doesn't view himself that way. "I do not see myself as the squeaky wheel type. Outwardly I am laid back and tend to do my own thing without asking for much until somebody puts an obstacle in my way. Then the gloves come off."

Breton started building his device anyway. He repurposed existing equipment at the EPA lab in Alexandria, Virginia, or figured out other ways to scrounge the things he needed. Snap-On, a Wisconsin company whose products include auto diagnostic equipment, lent Breton something called a five-gas analyzer, one of the most expensive parts he needed. Although some higher-ups in the EPA looked askance at Breton's work, people in the Alexandria lab were supportive. "The lab personnel were eager to help me do things that were not traditionally done because it was exciting times and we were breaking new ground," Breton said. Often he worked on the project nights and weekends, for no extra pay.

Some of the parts that Breton required didn't exist, so he had to invent them. For example, he devised a method to correct for normal fluctuations in the amount of exhaust coming out of a tailpipe, which make it hard to get consistent readings of pollution levels. He wrote the software that processed the data from the sensors and provided continuous readings of a car's emissions output. He figured out a way to connect the whole system to the vehicle's own diagnostic system.

The prototype looked like an improvisation, but it was fully functional. A pipe was connected to the exhaust. A sensor in the pipe collected data. A cable, wrapped in black protective foam, ran from the sensor through the car's open rear window and into the five-gas analyzer, a red metal canister the size of a large toolbox with a monitor screen in the side. The analyzer was propped up in the backseat, connected to a laptop computer that processed data and displayed the readings. A green rectangular box at the bottom right of the screen showed levels of NOx, or nitrogen oxides, one of the major pollutants associated with diesel.

The system proved its worth when Breton used it to double-check GM's attempt to fix the offending Cadillacs. GM's proposed repair worked in the lab, but made only a small, inadequate improvement in road performance. "They had to come up with a better fix that did work in the real world," Breton said.

Breton continued to refine the system, which he called ROVER, for Real-time On-road Vehicle Emissions Reporter. He added a GPS device so that it was possible to record where the vehicle was during tests and figure out the effect of terrain on the data. In 1997, the technology helped expose a device in Ford Econoline diesel vans that turned down emissions equipment at highway speeds. The defeat device improved fuel mileage but led to higher nitrogen oxide emissions. Ford paid $7.8 million to settle the violations and recalled the vans.

That same year, a technician in Ann Arbor noticed strange readings during lab tests of a heavy truck engine. After a certain period of time, the diesel engine's output of nitrogen oxide emissions suddenly doubled. Breton happened to overhear a colleague talking on the phone with someone in Ann Arbor about the engine's funny behavior. "When he got off the phone we discussed it," Breton recalled later. "I said something like, 'let's go get one and test it with ROVER.'" The Virginia Testing Laboratory rented a truck and hired a driver (Breton did not have a license to drive a heavy truck) and measured the vehicle's performance both in the lab and, with ROVER, on the road. Sure enough, the data from ROVER showed that nitrogen oxide emissions doubled when the driver shifted the manual transmission from sixth to seventh gear, as would be normal when accelerating to highway speeds. Breton suspected that the engine was programmed to adjust the timing of the ignition once the driver shifted into seventh gear.

The injection timing, which can be finely calibrated with the help of engine software, is a science unto itself. The goal is to squirt the fuel at the right moment to extract as much energy as possible while minimizing emissions. But it is often impossible to optimize both fuel economy—a crucial selling point in the truck industry—and emissions at the same time. Nitrogen oxides proliferate at the very high temperatures that occur inside diesel engines. When a diesel engine is running, the cylinders become hot enough to fuse nitrogen and oxygen from the atmosphere into NOx molecules with all their dire side effects. (The catalog bears repeating: asthma, chronic bronchitis, cancer, risk of heart attacks, and such environmental effects as smog formation, acid rain, and accelerated climate change.) Timing the fuel injection to achieve maximum efficiency may, as a side effect, raise the combustion temperatures inside the cylinder and increase production of nitrogen oxides.

Proving Breton's theory was difficult, though, because the engine's computer didn't provide data on ignition timing. Breton solved the problem by repurposing a circuit he had developed for a personal project, what he calls "a very sophisticated electronic bicycle trainer."

Somewhat to his surprise, it worked, showing that the engines were indeed programmed to ignite the fuel inside the cylinders a split second sooner when the car was in a higher gear.

A colleague of Breton's invoked the agency's regulatory authority and wrote to all the major truck and engine manufacturers asking if they were using engine control strategies that led to increased highway emissions. The people in Virginia did not initially tell EPA higher-ups what they were doing, fearful that industry-friendly forces inside the EPA would interfere. U.S. rules allow vehicle manufacturers to reduce emissions controls under certain conditions, for example, to protect the engine from damage. But manufacturers are required to report any so-called auxiliary engine control devices to the EPA and get permission to use them. If they don't, it's a defeat device.

The major truck engine manufacturers wrote back admitting that, indeed, they did use similar emissions "strategies." The truck makers had never been the target of a major compliance action before and, according to Breton, probably didn't realize what a serious offense they had committed. As Breton suspected, they had fiddled with the ignition timing without telling the government. Cummins Engine Co. Inc., for example, varied the timing of fuel injection to the cylinder, depending on whether the vehicle was on the highway or undergoing the federal test procedure.

Cummins denied that it did anything illegal, but along with other manufacturers eventually agreed to a settlement in 1998 that dwarfed anything the industry had seen before. All told, the settlement cost the companies $1 billion. That included a fine of $83.4 million, at the time the largest penalty ever for a violation of the Clean Air Act. Cummins and Caterpillar Inc. each paid $25 million of the fine. Smaller fines were imposed on the others: Detroit Diesel Corp., Mack Trucks Inc. and its partner Renault Vehicules Industriels, Navistar International Transportation Corp., and Volvo Truck Corp. In addition, the companies agreed to pay a total of $110 million to research and develop technologies to reduce nitrogen oxide emissions. And they promised to bring cleaner engines on the market, rebuild older engines to be cleaner, and recall some of the offending vehicles. The EPA estimated the total cost of the additional measures at $850 million.

The size of the truck settlement was a warning to the entire motor vehicle industry. From that point on, it was obvious that vehicle manufacturers caught using defeat devices were risking serious financial fallout.

The three cases also demonstrated the value of testing vehicle emissions on the road. No one argued that lab tests should be abolished. Controlled conditions were still needed to measure vehicles against regulatory standards. But road tests provided a reality check. Breton won numerous awards and commendations from the EPA and his enforcement work. The EPA later licensed some of Breton's patents to Horiba, a Japanese maker of measuring equipment. Horiba used the technology to produce so-called portable emissions-measuring systems, or PEMS, making the technology commercially available. Spot checks of diesel trucks using PEMS equipment became a standard part of regulatory practice in both the United States and Europe. Breton's patents earned royalties for the EPA, and, under government rules that apply to civil servant inventors, he also got a cut.

But Breton said he was punished for bypassing the normal chain of command in pursuing the truck makers. His career stalled, and he did not receive promotions after 1995, despite inventions that generated patent royalties for taxpayers, not to mention emissions detective work that allowed the government to collect tens of millions of dollars in fines.

Breton was not alone in believing that the EPA was more interested in writing regulations than in actual enforcement. An investigation by staff of the Commerce Committee of the U.S. House of Representatives accused the agency of being too close to the industry and ignoring warnings from experts going back to the early 1990s that modern engine technology made it easy for vehicle makers to circumvent clean air rules. The report portrayed the EPA as "a bureaucracy too slow and too arrogant to understand the profound changes taking place in emission control technology." Representative Tom Bliley, a Republican congressman from Virginia, accused the agency of being "all show and no go." It's worth noting that the truck makers' malfeasance took place during the presidency of Bill Clinton, whose vice president, Al Gore, later had a successful second career as an environmental crusader.

Though the testing of trucks on the road eventually became standard practice, there was little interest among government agencies in testing passenger cars on the road. From a regulatory point of view, there didn't seem to be enough reason to bother. Almost all cars sold in the United States ran on gasoline. In gasoline engines, the fuel burns at a lower temperature than in diesel engines, producing fewer nitrogen oxides. By the end of the 1990s, catalytic converters had developed to the point where they could effectively neutralize carbon monoxide, hydrocarbons, and other pollutants. To many people in the regulatory community, the problem of gasoline engine emissions was largely solved. Passenger cars with diesel engines were simply too rare in the United States to be on anybody's radar. Besides, many engineers remained unshaken in their belief that laboratory testing was superior. In 2001, the EPA decided to consolidate its emissions testing in Ann Arbor. The testing lab in Alexandria, where Leo Breton had built his ROVER, was closed.

In Europe, there were a lot more reasons to be scrutinizing diesel cars. Diesel was surging in popularity, in large part because of Volkswagen's success combining computer technology and fuel injection to produce diesels that did not rumble and spew clouds of dark smoke. By 2002, 40 percent of new car registrations in Western Europe were diesels, double the market share of a decade earlier. In Europe, where gasoline could easily cost four times as much as in the United States, the fuel economy of diesel was a powerful argument. A car could go about 15 percent farther on a tank of diesel. What's more, the car and truck makers had successfully convinced many European governments that diesel was more environmentally friendly in order to justify more favorable tax treatment. For one thing, it produced less carbon dioxide, the main cause of global warming. The industry downplayed the harmful effects of nitrogen oxides.

Environmental groups in Europe were already becoming skeptical about diesel's alleged benefits, and suspicious of claims that diesel was clean. The settlement with truck makers in the United States alerted activists to the risk that computer technology, now a standard component of vehicle engines, could be used to recognize when a car was on rollers in a regulators' lab—for example, by detecting when the wheels were moving but the steering wheel was not. In 1998, soon after the American truck makers had been exposed, the European Federation for Transport and Environment published a paper entitled "Cycle Beating and the EU Test Cycle for Cars." The author, Per Kageson, noted that European Union emissions tests were not very demanding or realistic. During the simulated driving cycle, cars were eased from zero to 50 kilometers per hour, or about 30 mph, in half a minute. That is a painfully slow rate of acceleration, one that would test the patience of even the pokiest driver. (A 1971 Volkswagen Beetle, not exactly a race car, could go from zero to 50 kph, in ten seconds.) More importantly, Kageson pointed out that the rigid and predictable nature of official tests was an open invitation for carmakers to cheat. "Car manufacturers can use modern electronic equipment to adapt the engine to any type of test cycle," Kageson wrote. "They can even tell the computer of the car how to recognize when the car is being driven according to a specific test cycle and adjust the combustion accordingly." The driving pattern used for tests should be at least partly random, Kageson argued, so that it would be more difficult for the car's computer to recognize and thwart it. It would take almost two decades before anyone heeded that advice.

The 1998 settlement between truck makers and the EPA had another consequence that did not seem important at the time but would later have enormous significance. Under the terms of the agreement, the truck makers were required to show that their new engines coming on the market were compliant with the latest regulations, on the road as well as in the lab. The manufacturers went looking for a contractor who could do the testing.

There weren't a lot of candidates. Emissions research is not a very sexy corner of the automotive business. People who choose to make automotive engineering a career are often people who love speed and horsepower. Emissions research doesn't offer that kind of gratification. On the contrary, emissions equipment tends to diminish performance, adding weight to vehicles and forcing manufacturers to make compromises in the way they tune engines. Emissions equipment is the annoying neighbor who shows up at your party and tells you to turn down the music.

One of the very few places that had made this unglamorous science its specialty was West Virginia University. Located in the small city of Morgantown, near the border with Pennsylvania, West Virginia University was a respectable but not especially prestigious outpost of academia. The state of West Virginia is famously hilly and rocky, inhospitable to both agriculture and manufacturing, helping to make it one of the poorest states in the United States. Morgantown has elements of a funky college town but also looks as if its heyday was about a century ago. It's a city where a visitor can eat in a restaurant selling local craft beers but, on the way home, encounter a homeless man sleeping under the marquee of the downtown's abandoned movie theater. Morgantown's most famous native son was Don Knotts, the comic actor who played Barney Fife, the excitable deputy sheriff on the 1960s sitcom *The Andy Griffith Show.*

West Virginia University, the state's flagship public institution, could not count on state tax revenue for much of its funding and had to scrounge elsewhere. That was also true of the Center for Alternative Fuels Engines and Emissions, or CAFEE, part of the university's College of Engineering and Mineral Resources, located a little north of the center of Morgantown on high ground overlooking the green waters of the Monongahela River.

The center had gotten its start in 1989 when the U.S. Department of Energy was promoting the use of natural gas for trucks, buses, and other heavy-duty vehicles. The Energy Department wanted to know what kind of emissions the alternative fuel produced—whether, in fact, it was cleaner. WVU, which already had a facility doing research in engine combustion, received a contract to develop an emissions-testing laboratory that could be hauled to different locations to test fleets of vehicles. Since it was impractical for municipalities to ship their buses or garbage trucks to a government lab, the lab needed to go to them.

The WVU engineers adapted existing testing equipment, putting it on wheels so that it could be pulled around the country, analyzing the emissions of garbage trucks or municipal buses that had been converted to run on natural gas. The equipment, which the students and professors built themselves, wasn't exactly compact. It included a platform with rollers, known as a dynamometer, big enough for a heavy-duty vehicle to sit on while its emissions were being tested. The dynamometer rode around the country aboard a forty-foot flatbed trailer pulled by a truck. A second truck hauled a portable laboratory, while a third hauled a trailer with tools and support equipment. Dan Carder, an undergraduate who earned tuition money working on the project, likened it to a traveling circus. "The only thing we were missing was the tent and the elephants," he said. The WVU crew took the mobile lab to both coasts, to Alaska, and as far south as Mexico City.

The mobile-testing project gave WVU a reputation in the industry and made it a strong contender when the truck makers were looking for a way to fulfill the terms they had agreed to after having been caught cheating on emissions. The engine makers had promised the EPA they would build cleaner engines; they needed independent confirmation that they were complying. Some WVU engineering graduates had gone on to Cummins, one of the truck makers, and knew what people at the university could do. Because CAFEE had done work both for regulators and for manufacturers, it had a reputation for objectivity. In 1999, CAFEE won a $1 million contract to do the compliance testing. It was just a sliver of the $1 billion truck makers' settlement, but it was a windfall by the standards of the emissions engineers at the university.

The money also came with a technological challenge. The truck makers' case had made it clear that laboratory emissions testing needed to be augmented by road testing. But there was no established procedure for conducting road tests of heavy trucks. Dan Carder, who by then had graduated from WVU and become a full-time staff member of CAFEE, was aware of Leo Breton's work and considered him the "grandfather" of on-road emissions testing. But the technology had to be adapted to provide the kind of data that the EPA was demanding from the truck makers.

The engineers at WVU had doubts about on-road testing, because of all the variables like weather and terrain that could skew the data. To Greg Thompson, an associate professor who worked closely with the staff at CAFEE, "it was more accurate than putting your thumb up." But the truckers' agreement with the government called for vehicles to be tested on the road, so Thompson, Carder, and others at the university set about building the best system they could with the technology available.

They had much experience inventing things and improvising solutions to problems. Their lab was in a cinder block building on campus originally used as a science lab for the university's agriculture program. It wasn't an ideal space for an emissions lab, and it took a lot of creativity to make it work. In an attic-like space above a chamber where engines were tested, Carder had welded much of the piping necessary to collect and analyze the exhaust gases, after first teaching himself to weld. He and the other members of the staff had to be mechanics as well as engineers and academics. To do stationary tests of truck engines, they had to know how to remove the motors from the vehicles and remount them on test beds inside the chamber. When they were done testing, they had to reinstall the motors in the trucks. Carder, who had an easygoing manner and a bit of a West Virginia drawl, was a natural for the job. A native of Parkersburg, a small city on the Ohio River about 110 miles west of Morgantown, he was a compulsive tinkerer. Given a bicycle for his birthday when he was a boy, Carder dismantled it before his birthday party was over. At sixteen, he restored a 1967 Mercury Cougar. And it wasn't just any Cougar, but one with an engine known as the 428 Super Cobra jet, designed for drag racing.

Thompson was more the academic. Tall, whereas Carder was compact, he had a straightforward, factual manner and a dry sense of humor. Thompson joked that he spent 80 percent of his time teaching and the other 50 percent overseeing the lab. He was the one who made sure that the work done by Carder and the grad students could be translated into papers that could be published, lending the operation more credibility. His doctorate in mechanical engineering from West Virginia University gave Thompson the credentials necessary for papers to be considered by scholarly publications. While Thompson taught courses like applied thermodynamics or machine design, Carder was the hands-on guy. Thompson often expressed good-humored frustration at how difficult it was to get Carder out of the lab long enough for him to finish his dissertation and earn a doctorate himself.

Carder and Thompson cobbled together a road testing system using the limited equipment then available for collecting and analyzing emissions underway. Carder called it a "bread box approach," swapping out sensors and other components from manufacturers like National Instruments and Horiba, and bumping the gear around on the serpentine West Virginia roads to test its durability. They improvised some of the necessary plumbing needed to capture the exhaust using off-the-shelf clamps and hoses. After a year of experimentation, they had what they called a mobile emissions measurement system, or MEMS.

Over the next seven years, working under the supervision of the EPA, the West Virginia team tested 170 heavy-duty diesel engines. Carder and Thompson tested not only tractor-trailers but also school buses, city buses, cargo vans, dump trucks, and cement mixers. They drove in temperatures as high as ninety degrees Fahrenheit, and they tested trucks in snowstorms when the temperature was twenty degrees. They compared the performance of newer engines to that of older ones. The research, completed in 2006, showed that, in fact, the truck makers had substantially cut emissions of nitrogen oxides to within legal limits. Carder and Thompson didn't see themselves as environmental crusaders. "We're just trying to advance the technology," Thompson said. "We're not there with any dog in the fight." But their work had helped make the air cleaner. Carder and Thompson and their staff and changing cast of graduate students had also advanced the science of on-the-road emissions testing.

Hardly anyone outside the industry was aware of their work, which they documented in publications such as the *Journal of the Air and Waste Management Association* and the *Journal of Commercial Vehicles*. Even many people in the auto industry were ignorant of the research conducted at the university, with its functional brick buildings and well-manicured lawns tucked amid the West Virginia hills. They did not realize the degree to which it had become possible to measure vehicle emissions under everyday driving conditions. Among the clueless, it later turned out, were the engineers in Wolfsburg.

CHAPTER 8

Impossible Doesn't Exist

BEING A MANUFACTURER of affordable cars for the masses was never going to be enough for someone of Ferdinand Piëch's ambition. So when Rolls-Royce Motor Cars, the British company that made Rolls-Royce and Bentley, came up for sale in 1997, it should not have come as a surprise—even though it did to many people in the industry—when Volkswagen joined the bidders. The idea was absurd at one level. It would be a marriage between a maker of cars defined by their accessibility to the *Volk* with an automaker defined by its inaccessibility. At the time, a union between Volkswagen and Rolls-Bentley may have made sense only to Piëch. After a wrestling match with BMW, which was also interested in the British luxury marques, a compromise was struck that left Volkswagen with the Bentley brand and BMW with Rolls-Royce.

While the rest of the auto industry was still puzzling over the logic behind the deal, Piëch went further. First, in July 1998, Volkswagen bought Bugatti, which was more of a myth than a carmaker. Bugatti had produced some legendary race cars and luxury sedans between the world wars. But after World War II it made vehicles sporadically and only in small numbers. It was bankrupt when Volkswagen bought it. Then, in September 1998, Piëch surprised again. Volkswagen bought Lamborghini, the ultimate in Italian super sports car machismo.

Bentley, Bugatti, and Lamborghini had much in common with each other, if not with Volkswagen. All three were glamorous and chronically unprofitable, founded by visionaries intent on testing the limits of motor vehicle luxury and performance. The bottom line was a secondary consideration. There was a distinct echo of Ferdinand Porsche.

Bentley, founded in 1919 by Walter Owen Bentley, who made a name for himself designing aircraft motors for the British armed forces during World War I, built cars that combined the luxury of a Rolls-Royce with the muscularity of a racer. Bentley had a single profitable year—1929—before Rolls-Royce acquired it out of bankruptcy in 1931. Bugatti's founder, Ettore Bugatti, who had studied art before teaching himself the principles of automotive engineering, built cars that were elegant, fast, and stupendously impractical. Bugatti's Type 41 Royale, first sold in 1926, had what for that era was an astonishing top speed of 165 kilometers per hour (about 100 mph). Its motor, modified from an aircraft design, had a cylinder displacement of almost thirteen liters, more than six times as large as that of a modern Passat sedan. There were no EPA mileage ratings at the time, but the Royale, made in Alsace, could not have gotten more than a few miles to the gallon. Reflecting Bugatti's art school training, the Royale was a beautiful car, with sculptural, one-piece fenders on each side that began over the front wheels, narrowed near the middle of the car, swept under the doors, then widened near the trunk to shield the rear tires. With a price of 160,000 marks ($38,000 in 1926), the Royale was also the most expensive car of its time. It was not an ideal car to bring on the market on the eve of the Great Depression. Only six were built. Of those only three were sold. Ettore Bugatti, who belonged to the same generation of auto pioneers as Ferdinand Porsche, died in 1947. The decades that followed saw occasional attempts to revive the name, but only a few hundred cars were produced. Bugatti lived on mostly as a label licensed to an eclectic collection of products with luxury pretensions, from umbrellas and leather goods to motorboats.

Ferruccio Lamborghini was the only one of this trio of automotive dreamers with a talent for making money, though not from sports cars.

Lamborghini made a fortune in the tractor business before deciding, in 1963, to realize his vision of the perfect *grand turismo*. Lamborghinis came close. Unapologetically macho—the Lamborghini trademark was a raging bull—they were essentially race cars that had been tamed just enough to be allowed on the road. The Miura, built beginning in 1966, stood barely a meter high to minimize wind resistance. It had a top speed of 278 kilometers per hour (about 170 mph). Customers bought as many as the factory in Sant'Agata Bolognese, in northern Italy, could produce. But only 150 were made from 1966 through 1969. Showered with accolades but losing money, Ferruccio Lamborghini sold out to a Swiss businessman who proved to be the first of a series of luckless owners, at one time including even the U.S. automaker Chrysler.

In theory the three brands would lend some of their cachet to Volkswagen through the so-called halo effect, enhancing the image of the proletarian main brand and making it easier to charge a higher price. The alleged benefits of this strategy were hard to quantify. At auto shows in Detroit and Geneva, the exotic brands no doubt helped draw visitors to the Volkswagen Group display. But among people who didn't go to auto shows it was unclear how many car buyers were even aware that Volkswagen owned Bentley, Bugatti, and Lamborghini, which were sold in separate showrooms. Some Volkswagen executives argued that the brands also provided platforms to try out costly new technologies, like the use of lightweight carbon fiber for body parts in place of steel, which would eventually make their way into less expensive vehicles. In addition, the chance to work on exotic cars may have helped Volkswagen attract top engineers and designers. Whatever the business logic, it was obvious that Piëch's personal fascination with the three carmakers played a big role in the acquisitions. He bought them, it seemed, because he could.

Piëch made no apologies. He even admitted that his interest in Bugatti was inspired in part by his youngest son, Gregor. While vacationing on the Spanish island of Majorca, Piëch showed the lad a model of a Rolls-Royce on display in a local shop that catered to tourists. Newspapers had just reported that Rolls and Bentley were for sale. But Gregor pointed to another model on display that he liked better than the Rolls: a model of an antique Bugatti. Piëch bought it for him. It was tempting to see the anecdote as a metaphor for Piëch's acquisition of Bugatti, Bentley, and Lamborghini. They were for sale. He liked them. He had the money to buy them, and so he did. At other companies there might have been dissent from within the supervisory board or from large shareholders about acquisitions that would be costly and offered questionable benefits. But no one stopped Piëch. The acquisitions demonstrated the nearly absolute control he had established over Volkswagen.

For Bentley, Bugatti, and Lamborghini, the benefits of being part of Volkswagen were clearer. Volkswagen provided the consistent financing they needed to develop new models, a deep pool of engineers they could draw on for the labor-intensive process of designing a new car, plus the procurement power of a huge corporation. Despite their exclusivity, the luxury brands discreetly borrowed Volkswagen parts in places that were invisible to customers.

Volkswagen soon signaled that the three brands were not just baubles; it was serious about reviving them. The company invested 1.1 billion deutsche marks, about $530 million, to upgrade Bentley's manufacturing operation in Crewe, a city located roughly halfway between Manchester and Birmingham, and to develop new products. The first model designed under Volkswagen ownership, a new version of the Arnage sedan, debuted in 2002. Equipped with a 456 horsepower V-8 motor, the Arnage could accelerate from zero to 100 kilometers per hour (60 mph), in 5.5 seconds. That was extremely fast for a full-size limousine weighing more than 5,000 pounds. The Continental GT, a two-door coupe, followed in 2003. Aimed at younger buyers, the GT had a twelve-cylinder motor with a top speed of 320 kilometers per hour (about 190 mph). The GT sold for 110,000 British pounds ($175,000), which in the luxury segment qualified as a bargain. It was a hit with wealthy young jet setters. Paris Hilton drove one. Before Volkswagen took over, Bentley sold fewer than one thousand vehicles a year. In 2007, it sold more than ten thousand and reported a profit of €155 million ($225 million).

Lamborghini, which operated as a division of Audi, provided an example of how an exclusive brand could draw on the resources of its parent. The chassis for the Lamborghini Gallardo was made at an Audi factory in Neckarsulm, Germany, then finished at the Lamborghini workshop in Sant'Agata Bolognese. The starting price in the United States was around $180,000—as super sports cars go, also a bargain. By 2007, sales had risen from a few hundred cars a year to 2,400, and the unit reported a profit of €47 million ($32 million).

Since Bugatti was not manufacturing any cars when Volkswagen bought it, the brand had little value in the automotive market outside aficionado circles. That was not Piëch's motivation, in any case. His aim was to position Bugatti in the stratosphere of the automotive market, as Ettore Bugatti had tried to do between the wars. Volkswagen went so far as to purchase the château in Molsheim, in the Alsace region of eastern France, where Ettore had once kept his workshops and lived in a grand villa (part of a lifestyle that helped explain the chronically perilous financial state of his company). Molsheim was revived as the workshop where Bugattis would be assembled by hand, while also serving as a customer service center and showcase for the brand. In an expression of how much importance Piëch attached to Bugatti, Piëch put one of his inner circle of favored engineers in charge: Karl-Heinz Neumann, nicknamed "Yogi," as in the cartoon character Yogi Bear. The nickname referred to Neumann's style of management and his way of barreling past obstacles to bring projects to completion. Still, the project was plagued with problems, and it wasn't until 2005 that Bugatti finally produced its first car under Volkswagen ownership, the Veyron. In the spirit of Ettore Bugatti, it was fabulously expensive and ridiculously overpowered. The Veyron's sixteen-cylinder motor produced 1,001 horsepower and was, Volkswagen boasted, the most powerful motor ever offered in a production passenger car. The Veyron could reach 60 mph in 2.5 seconds, and its top speed was around 250 mph. Michelin produced a tire specially for the Veyron because no commercially available ones could handle the velocity without disintegrating. The price was well over $1 million. The Veyron sold better than Bugatti's Type 41 Royale—eighty-one were delivered in 2007. Still, again in the spirit of Ettore, Bugatti managed to lose money.

The luxury brands served at least one purpose. They declared to the rest of the auto industry once and for all that Volkswagen was striving for a new identity beyond the confines of the overcrowded, marginally profitable mass market. Bentley, Lamborghini, and Bugatti were evidence that the company that had built one of the most practical cars in history, the Beetle, also had the engineering know-how as well as the ambition to build some of the most over-the-top vehicles ever to hit the road. There was more to come.

Piëch argued that the high-end brands allowed Volkswagen to hang on to customers as they became wealthier. In his view, Volkswagen needed to defend itself against BMW and Mercedes, which were moving down-market and encroaching on Volkswagen's turf. In 2001, for example, BMW launched an updated version of the Mini brand, which the company had acquired as part of an otherwise ill-fated takeover of the British carmaker Rover. The Mini was as affordable as a Golf, but more stylish and fun, with lively performance that BMW likened to driving a go-cart. The Mini opened up a new category—the hip, upscale small car. It was a hit, and a threat to Volkswagen.

Piëch decided to retaliate by building a car that would encroach on BMW and Mercedes, yet still wear a Volkswagen badge. A Passat driver who wanted to move into the luxury class should be able to do so without forsaking the Volkswagen brand. At least that was the theory. Thus was born the Phaeton, which like Bentley, Lamborghini, and Bugatti was designed to empty the thesaurus of superlatives yet, unlike the other exotic brands, still be a Volkswagen. Piëch envisioned the new model as a competitor to the top-of-the-line models from Mercedes and BMW, the S-Class and 7 Series. The Phaeton, however, would be even better, with features such as a compressed air shock absorption system, a nearly vibration-free twelve-cylinder motor, and a draft-free climate control system that cooled the car interior without blowing cold air on the passengers. (Later the Phaeton was available with a ten-cylinder diesel motor as well.)

What's more, the Phaeton would not be built in some drab industrial zone. The car would have its own showcase factory in Dresden, smack in the middle of downtown in a modernistic building with glass walls, a parquet floor, and its own sixty-five-seat gourmet restaurant. Volkswagen christened it the Gläserne Manufaktur, the Transparent Factory. Customers could watch their cars being assembled by workers in white overalls, then drive their new Phaeton home. It was auto manufacturing as performance art. "The entire assembly of the Phaeton is made into a public spectacle," Volkswagen said.

Piëch entrusted Phaeton development to one of his closest protégés, Martin Winterkorn, who had been head of quality control at Audi and was known for his attention to detail as well as his penchant for yelling when anything displeased him. After five years of development, Phaeton production began in 2001. The reviewers were impressed. A writer for *Car and Driver* found the Phaeton twelve-cylinder engine so quiet he wasn't sure that the key worked. "Only after sighting the tachometer needle trembling at 640 rpm do I realize the engine is alive," he wrote. But, despite the positive press, it quickly became apparent that people who could afford close to $100,000 for a car were not likely to buy a Volkswagen, no matter how refined, when they could have a Mercedes, a BMW, or an Audi for the same price. If anything, the Phaeton stole buyers from the Audi A8, with which the Phaeton shared many parts.

Sales were disappointing from the beginning, rarely reaching even half the annual target of 20,000 vehicles and nowhere near enough to justify the enormous investment in developing the car. The Transparent Factory alone cost €187 million ($168 million in the 2001 exchange rate). The Phaeton was also costly for owners. The twelve-cylinder motor was so tightly squeezed into the engine compartment that mechanics had to remove it completely to perform relatively simple repairs. It cost €5,000 just to replace the electric motor used to start the engine, according to *Spiegel Online*. High maintenance costs were one reason that Phaetons depreciated rapidly on the used car market, selling for less than €20,000 before the odometer had hit 100,000 kilometers, or about 60,000 miles. Years later a veteran executive who had worked closely with Piëch (and insisted on anonymity) bluntly described the whole Phaeton project as "insane." There is sometimes a fine line between genius and megalomania—in the eyes of many, Piëch crossed it with the Phaeton.

Piëch's acquisitions had established Volkswagen's ability to produce cars that pushed the limits of automotive engineering. The engineering excellence of the mass-market vehicles had also improved dramatically. By the end of the 1990s, there was little doubt that Piëch had turned Volkswagen around. Sales had nearly doubled by 1999, to €75 billion, about $75 billion. (The euro-dollar exchange rate was about 1 to 1 at the end of 1999). The company was still not very profitable, in part because Piëch plowed money into research and development and new factories. Volkswagen reported a net profit of €844 million in 1999, a return on sales of only a little more than 1 percent. But there had not been a loss since the one in 1993 that Piëch had effectively inherited from his predecessor.

BUT, AS SO OFTEN WITH PIËCH, there was a dark side to the story. He also acquired a reputation for pushing his engineers to the limit. Soon after Piëch took over Volkswagen, Paul A. Eisenstein, a veteran auto journalist, was among a group of reporters invited to Wolfsburg to view the prototype of a new sedan. Piëch told the journalists that the new model would leapfrog all competitors, and listed the new features and technical breakthroughs he planned to achieve. Someone asked what would happen if the engineering team couldn't deliver. According to Eisenstein, Piëch replied, "Then I will tell them they are all fired and I will bring in a new team. And if they tell me they can't do it, I will fire them, too."

Other people told similar anecdotes. Bob Lutz, former president and chief operating officer of Chrysler, recalled a conversation he had with Piëch in the 1990s. Lutz encountered Piëch at a dinner sponsored by the German Automobile Manufacturers Association, known by its German initials, VDA. Lutz, who speaks German, told Piëch how much he admired the exceptionally narrow body gaps in the latest version of the Golf, which had improved because of Piëch's obsession with the spaces between exterior surfaces like the doors, hood, and fenders. "I said, 'These body gaps are terrific, Herr Piëch. I wish we could do that at Chrysler.' " According to Lutz, Piëch then confided his recipe for achieving manufacturing excellence. He called the top managers responsible for engineering the car bodies and stamping the sheet metal into his office. He told them that he was tired of the lousy body gaps, and gave them six weeks to improve. As Lutz related the story, Piëch then informed the managers that he knew all their names, and if they failed to achieve perfect body gaps, they would be fired. Then he thanked them and told them to get back to work. When Lutz—who had been struggling to improve the body gaps at Chrysler—said he doubted that the same technique would work in the United States, Piëch told him that he was weak.

Lutz called Piëch "probably the greatest automotive product genius in the history of this industry." But he added, "I would not have cared to work for him."

Piëch was not a screamer. Another former executive (who requested anonymity for fear of retaliation) recalled an occasion when Piëch dreamed up a particularly difficult engine and transmission configuration for one of Volkswagen's models. When one of the engineers protested that it would be nearly impossible to make the configuration work, Piëch replied coolly, "If you don't want to do it, you don't have to." The implication was clear: the engineer was free to leave the company anytime he wished. Failure to achieve a technical assignment that Piëch had decided was within the realm of possibility was not an option. Around Volkswagen, there was a saying: "Geht nicht, gibt's nicht." Roughly translated, that means, "Impossible doesn't exist."

Piëch was a master of cutting down subordinates with a few icy words. Another former executive remembers attending a meeting where a colleague was making a presentation. Piëch grew bored and took an apple from a bowl of fruit on the table in front of him. He began to peel the apple with a knife. When he was done peeling the apple, the executive recalled, Mr. Piëch looked up and said, "Why is this guy still talking? Can't anyone tell this guy to just leave?" (Asked about these anecdotes, a lawyer for Piëch, Matthias Prinz, declined to comment except to say the stories did not correspond to the Piëch he knew.) On other occasions, according to several former executives, Piëch would sit through meetings without saying anything at all. Nor would his facial expression betray his thoughts. At the end of a meeting, when everyone looked expectantly for his reaction, Piëch would get up and leave without a word. He liked to keep people off balance.

Ever the engineer, Piëch took an obsessive interest in product minutiae and insisted that his top managers do the same. Very little escaped his gaze, and his word was law. Ferdinand Dudenhöffer, a former employee of Porsche who later became a university professor and prominent commentator on the auto industry, recalled one day when Piëch was in Salzburg and noticed that Volkswagens on display at the family-owned dealership were covered in snow. Piech was not pleased. Ever after, the story goes, no flake of snow was allowed to linger on any Golf or Passat at the lot in Salzburg.

Piëch also personally inspected and approved cars that would be displayed at auto shows. In November 1999, Piëch came to scrutinize models being prepared in Wolfsburg for the Detroit International Auto Show, which would take place the following January. He arrived with an entourage that included several members of the management board, some less senior executives, as well as engineers and technicians, primarily from research and development and from production. Auto shows are important marketing venues for carmakers. Still, the commitment of so much highly paid management talent for car show preparations was typical of Piëch's obsession with detail and his intense personal focus on products.

The person in charge of getting the cars ready in Wolfsburg before they were shipped to Detroit was a veteran executive named Walter H. Groth. Among the cars to be displayed was the sporty GTI version of the Golf, which was doing well in the United States. Groth had ordered a red one for the show with a black interior. It was a risky move on

Groth's part. As he well knew, Piëch did not like red cars with black interiors, which he said looked like coal boxes. Piëch preferred gray interiors. The problem was that the American public disagreed. Americans liked red GTIs with black interiors, no matter what Piëch thought.

Groth later recalled that, when Piëch saw the red-and-black GTI, he demanded, "Who ordered this car?" According to Groth, all fingers pointed at him. "I did," Groth replied.

"You are aware of my guideline that red GTI's cannot have an all-black interior?" Piëch said, according to Groth's account.

"The gray interior doesn't sell," Groth replied. "I ordered the car with all-black interior since it makes sense for our American market."

"Next time you order the car based on my guideline. Is that clear?" Piëch said. Turning to the others, he added, "And if he does it again, you are not going to build the car!"

According to Groth, though, Piëch was smiling as he spoke—and not in a sinister way. Groth did not get fired. "Piëch was not angry," Groth recalled later. "He doesn't show emotions. He always looks calm and speaks with a very quiet voice. He never yells. My point is that you can speak up and take a stand even in front of Dr. Piëch. However, you have to really know what you are talking about. You have to understand cars and engineering."

To Groth, Piëch's power depended not only on his ability to intimidate people but also on their willingness to be intimidated. "When it comes to leadership, I don't agree with how Dr. Piëch behaved on many occasions," Groth says. "And yes, in many people he instills fear . . . a lot of fear. But hey, it takes always two to tango."

Most executives, it was clear, were not ready to stand up to Piëch. On the contrary, his style of management emboldened subordinates to behave the same way toward their underlings. That is how corporate cultures—the unwritten rules that govern behavior inside a large organization—come into being. The people at the top set an example, and the people below them follow. One of the most assiduous practitioners was Winterkorn, who was as demanding as Piëch and one of his favorites, but physically more imposing, with a barrel chest and a megaphone voice. Winterkorn was known as a yeller with a short fuse. Arndt Ellinghorst, a management trainee at Volkswagen in the late 1990s, witnessed an episode where Volkswagen technicians were showing Winterkorn the infotainment system on the Phaeton, which was then in development.

Winterkorn mistakenly thought the push-button console was a touch screen, and became angry when it didn't respond. When the technicians tried to explain it to him, he accused them of treating him as if he were stupid. Ellinghorst, who later became an automotive industry analyst at Evercore ISI, an investment advisory firm, said he decided not to stay at Volkswagen because of its authoritarian culture, where employees were expected to simply follow orders. "VW was like North Korea without labor camps," Ellinghorst said, paraphrasing a description of the company once made by *Der Spiegel* magazine: "You have to obey everyone."

A few outsiders raised questions about Piëch's dictatorial management style. "The danger with Piech is that his personal ambition could cause him to overreach," *BusinessWeek* magazine wrote in 1998. "Moreover, his iron grip on VW means there are few checks and balances on his decisions." But as long as Volkswagen continued to grow and create jobs, there was no serious threat to Piëch's dominance.

Piëch stepped down as chief executive of Volkswagen in 2002, the year he turned sixty-five. Without question, Volkswagen had become a bigger, stronger company in the ten years since he had taken the top job. Sales had doubled to €87 billion. The number of cars produced had increased from 3.5 million to 5 million. The number of workers had risen from 274,000 to 325,000. In 2002, about half the employees worked overseas, up from 40 percent of the workforce a decade earlier. The company had also broadened its customer base and was able to appeal to more-affluent customers as well as younger people buying their first cars. In 1992, Volkswagen was overly dependent on the Golf, which sold almost three times as well as the more expensive Passat. By 2002, Passat sales had pulled almost even with those of the Golf. The Polo, which had become Volkswagen's entry level vehicle, was close behind. The more balanced product portfolio helped protect Volkswagen from shifts in demand.

The profit in 2002 of €2.6 billion ($2.7 billion) was an improvement over 1992, a recession year when the company barely broke even (and went on to record a huge loss the following year). But earnings in 2002 amounted to only 3 percent of sales, still a very meager return. And that was the second-best year in company history. Productivity had also improved during Piëch's tenure, but not radically. When Piëch first took up residence atop the brick office tower in Wolfsburg, Volkswagen produced an average of 12.8 vehicles per worker. By 2002, the ratio had risen to 15.4 vehicles per worker, a 20 percent increase in productivity. But Volkswagen still compared unfavorably with Toyota by most measures of profit and efficiency. The Japanese carmaker earned a 5 percent return on sales in the fiscal year ending March 2003, while the 235,000 employees in Toyota's automotive division made 6.1 million cars. That worked out to 26 vehicles per worker, 40 percent more than Volkswagen. Piëch had set out to close the productivity and profitability gap with Toyota, but had fallen far short. One is tempted to say that, by the standards he applied to subordinates who missed their targets, he should have been fired.

Porsche, on the other hand, had undergone a spectacular transformation, thanks in large part to a deal with Volkswagen. In a rare lapse by Piëch, Volkswagen had been late to exploit the boom in sport utility vehicles that began in the United States and was spreading to Europe and other parts of the globe. The Porsche chief executive, Wendelin Wiedeking, meanwhile, saw an opportunity for Porsche in the SUV market. A luxury four-wheel drive vehicle with Porsche driving performance and Porsche amenities would allow the company to offer another product to its affluent customer base. Owners of 911s would have a vehicle that could carry the whole family, but was still a Porsche. An SUV would also help Porsche expand into emerging markets like China, where there were enough newly affluent consumers able to afford a Porsche, but where the roads were often ill-suited to a ground-hugging 911.

In 2000, Porsche and Volkswagen agreed to develop the new SUVs together. Volkswagen's version was called the Touareg, while Porsche's was dubbed the Cayenne. There were significant cosmetic differences. The Cayenne's front hood sloped between the protruding headlights, a Porsche design trademark. But both vehicles were made at a vast factory on a plain outside Bratislava, Slovakia, that Volkswagen had bought after the fall of communism and expanded. The factory was notable for an aerial tramway, made by a company that also manufactures ski lifts, which transported finished vehicles from the assembly line to a test track located on the other side of a highway. The partially assembled Cayennes, already painted, were shipped to a spotless new Porsche factory and service center in Leipzig, where workers in red overalls and white T-shirts completed the assembly and customized the cars according to the wishes of buyers. Gold-plated gearshift knobs were not unheard-of. Purchasers could fetch their cars at the Leipzig factory and test them out at the adjacent racetrack, which replicated sections of famous Formula One courses.

The SUV cooperation was a good deal for Porsche. The up-front investment to develop the car and set up an assembly line was a fraction of what it would have been if Porsche had done so alone. And while the risk to Porsche was low, the profits were immense. In August 2002, Gerhard Schröder, the German chancellor, tightened the final screw on the first Cayenne to come off the assembly line in Leipzig. Less than two years later, the Cayenne accounted for almost half of Porsche's annual production of about 80,000 vehicles. Profit in the fiscal year ending on July 31, 2004, was €612 million ($832 million), more than double what it had been three years earlier. The 10 percent profit margin was better than Toyota's. In fact, Porsche became the most profitable carmaker in the world. Wiedeking, whose contract entitled him to 1 percent of profits, became one of the best-paid chief executives in Germany, and one of the most admired.

Wiedeking was a big personality, with a temper but also a hearty laugh and a taste for luxury, as well as a few eccentricities. Portly, with large eyeglasses, he had an almost childlike fascination with mechanical things. His hobbies included collecting model trains and antique tractors. At the dinner that Porsche staged every year at the Geneva Motor Show for journalists who cover the auto industry, Wiedeking held court at a large round table in a restaurant booked for the occasion, chortling at his own jokes. He would stay up until the early morning hours, drinking brandy and smoking cigars with any reporters who had the stamina to keep pace.

Wiedeking made a lot of people rich, not only the Porsche and Piëch families. While the family retained all of the sports car company's voting shares, half the equity was in the form of preferred shares that had no votes but entitled the owners to a share of the profits. The preferred shares traded on the German stock exchange and attracted a broad spectrum of investors. Under Wiedeking, the value of both the preferred and the voting shares soared more than 1,000 percent.

Less certain was whether the Touareg-Cayenne cooperation was as good a deal for Volkswagen as it was for Porsche. The sports car company seemed to be reaping a disproportionate share of the profits, while the partners in Wolfsburg took most of the risk. There was an obvious conflict of interest in Ferdinand Piëch's role as chairman of Volkswagen and his family's ownership of Porsche. Porsche paid €780 million ($1 billion) to Volkswagen for Cayenne chassis, bodies, and motors in its 2005–2006 fiscal year. That worked out to about €22,000 ($30,000) per vehicle, or less than half the Cayenne's starting price of about €50,000 ($68,000). The arrangement left Porsche with plenty of room for a fat profit margin, especially considering that, with options, the price of a Cayenne could easily top €100,000 ($136,000). There were few complaints as long as both companies were profitable. Events soon exposed how fearful Porsche was that Volkswagen would come under control of people with less reverence for the historical ties between the two companies.

THE DECISION ABOUT who would succeed Piëch as his tenure came to an end legally belonged to the Volkswagen supervisory board. But Piëch had little trouble pushing through his choice: Bernd Pischetsrieder, the former chief executive of BMW. Pischetsrieder, who wore a neatly trimmed black goatee and was fond of Cuban cigars, had resigned from BMW in 1999 after the Munich-based automaker's acquisition of Rover proved to be a major financial drain. Soon afterward, Piëch, who had gotten to know Pischetsrieder when both were competing to buy Rolls-Royce, hired him to oversee quality control at Volkswagen as well as the division that produced SEAT brand cars in Spain. Pischetsrieder satisfied Piëch's expectations, allowing him to edge out other candidates for the chief executive post like Winterkorn, who became head of the Audi division.

Pischetsrieder's appointment as Piëch's successor did not, however, mean that Piëch's dominance of Volkswagen was coming to an end—not at all. Piëch was elected chairman of the Volkswagen supervisory board, an ostensibly part-time job that allowed him to keep a close eye on his successor. Under German corporate law, the supervisory board oversees the chief executive (known formally as the chairman of the management board) and has the power to remove him. Corporate governance experts frown on chief executives' becoming supervisory board chairmen after they retire. The risk is too high that the chairman will try to protect his legacy and block needed changes. The risk was perhaps particularly high in Piëch's case. He was unlikely to be a passive chairman. Pischetsrieder would need to stay on Piëch's good side if he hoped to keep his job.

Characteristically, Piëch chose to close out his tenure with a show of technical tour de force. Having been the first company to mass-produce a car that could travel 100 kilometers on three liters of fuel, Piëch ordered his developers to make the quantum leap to a car that could travel that distance on just one liter of fuel. The result after three years of development was a two-seat prototype called simply the *Einliterauto*—the one-liter car. With a streamlined lightweight body made of carbon fiber wrapped around a magnesium frame, and a Plexiglas canopy covering the driver, the car looked like a fighter jet minus the wings. The *Einliterauto* had cameras instead of rearview mirrors to minimize drag, and the wheels were lightweight titanium. The one-cylinder motor was a direct injection diesel, of course. The car reflected Volkswagen's emerging image of itself as a pioneer in environmentally friendly transportation. The *Einliterauto* demonstrated that Volkswagen's technical prowess could be deployed in the name not only of speed and horsepower but also of efficiency.

On April 15, 2002, Piëch drove the prototype from Wolfsburg to Hamburg to attend the Volkswagen annual meeting, and for the first time to show the world what the *Einliterauto* could do. To save a few more grams, Piëch even wore a special pair of lightweight driving shoes made by Ferrari. It was a rainy day, not ideal for maximum fuel economy. But when the car reached Hamburg, a journey of 237 kilometers, it had consumed an average of 0.89 liters per 100 kilometers, well under the goal. Piëch, chilled to the bone because he had not turned on the *Einliterauto*'s heat during the trip to save fuel, sank into a hot bath at the Vier Jahreszeiten, one of Hamburg's best hotels, satisfied.

The next day Piëch and Pischetsrieder drove together in the *Einliterauto* to the site of the annual meeting, a convention center in Hamburg where the leadership baton would be passed. Pischetsrieder squeezed into the passenger seat, which was located behind the driver as in a fighter jet. Piëch was at the wheel when the car pulled up to the convention center, where an entourage waited outside. Inside, about 3,500 shareholders—many of them elderly, dressed as if for church—lined up at buffet tables to collect free sausage and soup, an obligatory feature of German annual meetings known as the wurst dividend. There was some grumbling about Piëch's high-end acquisitions and whether they made any sense, according to *Automobilwoche*, a German trade paper. A groan went around the convention center when Pischetsrieder gave a pessimistic outlook for the year ahead, noting that sales in Germany so far that year were down. The Volkswagen stock price did not offer much consolation. At around forty euros, the price of Volkswagen preferred shares, the most widely traded, had quadrupled since Piëch took office. But the shares were down from a high of more than seventy euros in 1998. There were rumors that the depressed share price made Volkswagen a takeover target.

All in all, though, shareholders appreciated Piëch's role in averting disaster in 1992 and gave him a standing ovation. It is unlikely that any of them concerned themselves with Piëch's dictatorial style of management, or the fact that he had taken office complaining about the authoritarians in Wolfsburg, and left as the most authoritarian of them all. The important thing was the result. Piëch later wrote that the accolades, the sustained applause, and the praise from Klaus Liesen, the outgoing supervisory board chairman, left him cold. The drive in the *Einliterauto* was more interesting, he said. The comment was revealing, an acknowledgment that machines meant more to him than people.

Piëch claimed that he would not interfere with the daily management of Volkswagen. He sketched a vision of a quiet retirement, spending more time with his family and sailing on the high seas aboard a custom-made aluminum yacht fitted with technology that made it possible for Ursula and him to sail it alone. A 2002 photo in Piëch's autobiography even showed him peering over his bifocals from a bench in Salzburg, wearing a rumpled flannel shirt and trousers, holding a penknife and a stick of wood—just another pensioner enjoying the twilight of his life. No one who knew Piëch really believed that he was going to spend the coming years whittling.

CHAPTER 9

Labor Relations

PISCHETSRIEDER IMMEDIATELY SET ABOUT trying to change Volkswagen culture. He had spent most of his career at BMW and was not indoctrinated in the Volkswagen way of doing things. BMW, with its roots in fun-loving Bavaria, was a more decentralized, less authoritarian company than Volkswagen. Its top executives were regarded as mortals rather than gods, and it was possible for underlings to challenge their views. Pischetsrieder tried to relax Volkswagen's top-down way of doing things and to get employees to take more responsibility and initiative. It was not easy. One former department head recalled how Pischetsrieder came for a visit soon after taking over and asked employees what goals they had for their unit. The employees, who had never been asked for their opinions before, sat in stunned silence.

Pischetsrieder did not have a free hand to overhaul the management board, though. Piëch's position as chairman of the supervisory board made it difficult for Pischetsrieder to push out his predecessor's loyalists and install his own team. Some members of the management board had their own power bases. Among them was Peter Hartz, the management board member responsible for human resources. Keeping workers happy was always delicate work, and Hartz, who had negotiated the four-day workweek in the 1990s, was considered one of the best in the business at managing the relationship.

Hartz's stature was such that Gerhard Schröder, the German chancellor, chose him in 2002 to lead a special commission assigned to figure out how to reform German labor laws. It was a politically treacherous mission. Like many countries in Europe, Germany made it very difficult for companies to dismiss employees whose work was unsatisfactory, or to lay off workers when business was bad. The job protection rules meant that firms were very cautious about hiring people in the first place, because it would be so difficult to fire them if things didn't work out. In addition, unemployment benefits were generous enough that some people preferred to live on government support rather than work. Economists blamed Germany's high unemployment rate—11 percent in 2002—partly on these labor laws, because they discouraged hiring and put insufficient pressure on unemployed people to find jobs or acquire new skills. Schröder realized that changes were needed. But he faced strong resistance from the left wing of his own party, as well as from ordinary Germans who feared that they would be fired if job protections were relaxed. Before becoming chancellor, Schröder had been prime minister of Lower Saxony and a member of the Volkswagen supervisory board. He knew Hartz. A clever politician, Schröder realized that Hartz's working-class roots and track record dealing with Volkswagen workers made him an ideal person to find a compromise.

Schröder's faith in Hartz proved to be well founded. The program worked out by the Hartz Commission, as it came to be known, loosened some restrictions on hiring and firing and made it easier for companies to hire temporary workers or part-time employees who were exempt from most job protections. The program also set a time limit on unemployment benefits. After a maximum of two years—and as little as six months for younger people—jobless people received a subsistence allowance that came to less than €400 ($540) a month for a single person. The relatively meager welfare payment, instead of open-ended unemployment benefits based on the person's last paycheck, created a stronger incentive for jobless people to accept lower-paying jobs. The program, approved by the Bundestag after bitter resistance from left-wing legislators, was by most measures a success. In the years to follow, the German unemployment rate plummeted. Making it easier to fire made it easier to hire. By 2016, Germany's jobless rate was below 5 percent and lower than that of the United States. But Hartz, who kept his job after Piëch turned over operational management of Volkswagen to Pischetsrieder, would not go down in history as a hero who put more than two million people back to work.

Hartz's downfall may have begun at an establishment in Prague known as the K5 Relax Club. The club describes itself as "an oasis of entertainment and relaxation" where "everyone will find what he is looking for." The small sign outside does not hint at the opulent, if gaudy, interior spread over four floors of an inconspicuous prewar building in downtown Prague. The management of the club insists, however, that it is not a bordello. What goes on between male patrons and the women who work the club's bar is their business, the managers say. The club simply makes rooms available. Many of the private chambers have themes, like an erotic amusement park. There is the Space Room, where customers may "feel like a cosmonaut, accompanied by Barbarella in the endless vastness of space." Or the Igloo, where a giant stuffed polar bear looms over a bed surrounded by a wall of fake snow blocks. "Experience how bodies on bodies will never let you freeze," the club advertises. Patrons may also engage the services of a Brazilian woman who lies on a table covered in nothing but fruit, a human buffet. She is popular with stag parties.

Among the club's regular visitors was Helmuth Schuster, the head of personnel at Skoda, the Czech automaker that Volkswagen bought under Hahn's leadership after the fall of the Czech communist government. Skoda combined its own body designs with Volkswagen engines and chassis and sold the vehicles at prices affordable to Eastern Europeans. Schuster was also a close confidant of Hartz's. In June 2005, reports appeared in German newspapers that Schuster had been summarily fired by Skoda. Details about the reasons for his departure soon began to leak out. Internal auditors had uncovered evidence that Schuster, a portly man in his midfifties, had taken a €100,000 bribe in India to influence the location of a new factory. He used the money, equivalent to about $120,000 at the time, to buy a Lamborghini, German newspapers reported. Another allegation was even more serious, raising questions about the whole codetermination system and the mythos surrounding worker-manager cooperation at Volkswagen. According to criminal charges later filed against Schuster, he had brought union officials and workers council representatives with him to the K5 Relax and other sex clubs, and charged Volkswagen for the cost. The so-called *Lustreisen*—lust tours—were part of a system designed to keep Volkswagen labor leaders happy and compliant.

EVIDENCE SOON EMERGED that implicated Hartz and Klaus Volkert, the head of the Volkswagen workers council. Volkert, one of the most powerful labor leaders in Germany, had been treated like a de facto member of Volkswagen management after becoming president of the *Betriebsrat* in 1990. He had a parking spot in the lot at Wolfsburg reserved for top executives and flew first class. In June 2005, Volkert unexpectedly resigned, in what the workers council initially claimed was part of a planned handover to his top deputy, Bernd Osterloh. But soon accusations emerged that Volkert had received millions of euros in "special bonuses" from Volkswagen. The company had also paid for Volkert's mistress, a Brazilian woman, to make regular trips to Germany for trysts. Hartz had approved at least some of the payments.

A central figure in the widening scandal was a Hartz deputy named Klaus-Joachim Gebauer, who was in charge of relations with the workers council. According to later court testimony, Gebauer had control over a slush fund, code-named 1860, used to pay for bordello visits and to provide prostitutes for union officials during trips outside of Wolfsburg. If the women did not appear soon enough, Volkert was said to have asked Gebauer impatiently, "Wo bleiben die Weiber?"—Where are the women? It became the German news media's most-quoted phrase of the scandal. Gebauer also had a lover who allegedly was put on the Skoda payroll, earning €50,000 over a twenty-month period, while not doing any company work. The German media feasted on the scandal, and it received extensive coverage abroad. But the affair had little if any effect on sales. Who cared what Volkswagen executives did in their free time as long as they made good cars?

Hartz resigned in July 2005. In a statement, he admitted no guilt, but said he took responsibility for what had occurred. "It's about more than just me as a person," he said. "It's about Volkswagen's reputation, about which I feel a special duty." Hartz also sought to remind people of all the good things he had accomplished. "I have invested all my energy to build and preserve competitive jobs at our company," he said.

PERHAPS NAÏVELY—or perhaps in a deliberate attempt to undermine Piëch—Pischetsrieder ordered a thorough investigation of the allegations. He hired the accounting firm KPMG to conduct an internal inquiry and formally requested an investigation by German prosecutors. In November, half a year after the scandal came to light, Volkswagen disclosed the preliminary findings of the KPMG inquiry. The auditors had found evidence that money had flowed into private accounts controlled by Schuster and Gebauer in connection with the factory in India. The audit also turned up evidence that the two men secretly owned firms that sold services to Volkswagen, including a company that was supposed to set up a VW and Skoda sales network in Angola. Gebauer had paid internal vouchers worth €939,000 ($1.1 million) to himself over the previous five years for purposes that included "travel, jewelry and visits to bars." The audit also confirmed that a "female acquaintance of Volkert"—a reference to the labor leader's Brazilian lover—had received €635,000 ($760,000). Some of the invoices submitted by her "were approved for payment by Hartz," the statement said. Volkswagen put the total financial damage to the company at €5 million ($6 million).

The obvious question was: What, if anything, did Piëch know? The corruption had begun years earlier under his watch, and there was clear evidence that Hartz, who had been a member of the management board alongside Piëch, was deeply involved. The sums of money involved were substantial. But no documents or other direct evidence emerged that linked Piëch to any of the payments. Hartz insisted that he acted without Piëch's knowledge.

Piëch gave his version of events on March 29, 2006, when he submitted to questioning from two state prosecutors in Braunschweig and two detectives of the Lower Saxony Landeskriminalamt, the state agency that conducts criminal investigations. Piëch insisted that he was unaware of large payments to Gebauer for travel and entertainment costs, which had bypassed normal approval procedures. "Anyone who did such a thing should be thrown out," Piëch said in a signed statement, which has not been made public before. Piëch allowed that he had heard rumors that Volkert had a Brazilian girlfriend, but he said he had not taken part in any discussions about the salary she received from Volkswagen, or the apartment she kept in Braunschweig at company expense.

Piëch conceded, though, that he deliberately kept his distance from financial transactions. "I never distributed money," Piëch told the investigators. "Rather, in these unpleasant matters I stayed out of the loop and delegated to others." The statement could be interpreted as an admission by Piëch that he turned a blind eye to such things and left the dirty work to others.

Piëch was called to testify in January 2008 during the trial of Volkert and Gebauer. By then seventy, he arrived at the courthouse in Braunschweig in a black Tiguan SUV driven by Ursula. Piëch allowed that he might have told the labor leader that he deserved better pay in recognition of his responsibilities, which were comparable to those of a management board member. But he denied knowing anything about payments to Volkert or bordello visits. If he had been aware of the corruption, Piëch testified, he would have objected vehemently. Prosecutors, based in Braunschweig, a city adjacent to Wolfsburg, did not pursue charges.

To this day, there is speculation about whether strings were pulled to protect Piëch. But it would have been difficult to convict Piëch of any crime without an incriminating document or other smoking gun linking him to the corruption. Hartz's insistence that Piëch was ignorant of the wrongdoing might well have created enough doubt to ensure acquittal in a criminal trial. But even if Piëch was as oblivious as he claimed, the scandal highlighted a woeful lack of financial and ethical oversight at Volkswagen. By Volkswagen's own estimate, a small group of high-ranking executives had spent five million euros without anyone's noticing. Coming after the López affair a decade earlier, the prostitution scandal served as a further warning that something was badly amiss in Volkswagen's corporate culture—in the norms that top managers believed applied to them. Piëch, still chairman of the supervisory board (his plans to spend his retirement sailing had never amounted to much), had been the dominant figure in the company during both scandals. It is fair to assign him the moral responsibility for these failings.

Pischetsrieder recognized that the scandal was in part the result of a dysfunctional management culture and pushed for reform, according to people who worked with him. The company announced that, to prevent future wrongdoing, it would hire ombudsmen to whom employees could report corruption on a confidential basis. "We are drawing far-reaching conclusions from this," Pischetsrieder said in a statement in conjunction with the report. "Volkswagen will be a more transparent company both internally and externally."

Piëch escaped any legal consequences, but the others were not so fortunate. Hartz admitted in court that he had funneled money to Volkert from 1995 to 2004. In 2007, he received a suspended two-year sentence and a fine of €576,000 ($780,000). Volkert suffered the most serious punishment, a prison sentence of two years and nine months. He served a year and nine months in a facility near Hanover where prisoners were allowed to leave during the day. Gebauer, who cooperated with prosecutors, received a one-year suspended sentence. Schuster, the former Skoda head of personnel, received fifteen months' probation and a €15,000 ($20,000) fine. At least by United States standards for white-collar crime, these were relatively mild penalties.

Hartz's humiliation did not end with his conviction. He became a focal point of public resentment toward the labor law reforms he had helped design. Despite the reforms' success in cutting German unemployment, many Germans perceived them as a loss of privileges. Hartz's name became synonymous with the subsistence benefits that jobless people received when their regular unemployment benefits expired. A German person receiving welfare was said to be "on Hartz."

The scandal unfolded at a critical moment in Volkswagen history, though few outsiders realized it at the time. A group of engineers in Wolfsburg was beginning work on a new engine known internally as the EA 189. EA stood for *Entwicklungs Auftrag*, or Development Order. The routine designation provided no hint of the project's importance. The new engine would be the standard diesel power plant for Volkswagen's fleet of small and midsize passenger cars, installed in millions of vehicles. The EA 189 would also mark a transition to the latest diesel technology, known as common rail. The new technology used a single reservoir of pressurized fuel—a tube-like "common rail"—to supply all the cylinders. With previous systems, each cylinder had a separate fuel supply. Common rail ensured more consistent delivery of diesel fuel to the cylinders, and it was also cheaper.

Development of the engine was doubly important because of the role envisioned for it in the United States. The two-liter, four-cylinder version would be installed in the Golf, the Jetta, the New Beetle, the Passat, and the Audi A3 as part of a plan to spread the gospel of diesel to Americans. Volkswagen had already turned its mastery of diesel technology to achieve dominance in Europe and marginalize rivals like General Motors' Opel division, which was slow to recognize how important diesel would become in Europe. "The market shares over the years changed drastically, mostly on the back of diesel," recalled David J. Herman, the former chairman of Opel. Now Volkswagen would try to use the same strategy to gain ground against Toyota in America, where hardly any passenger cars were diesels. Volkswagen would pitch the EA 189 motor as an alternative to the Prius, Toyota's pioneering hybrid. Volkswagen diesels would offer almost as good fuel economy as the Prius, but with sportier acceleration. So important was the EA 189 to Volkswagen's strategy in the United States that the company developed a special variant known as "the U.S. Motor."

The engineers designing the EA 189 worked in a sprawling office and laboratory complex west of the main factory building. The Volkswagenwerk had grown enormously since the days when the factory churned out nothing but Beetles. Wolfsburg had changed, too, since its Nazi beginnings as the awkwardly named City of the Strength through Joy Car. To the east of the factory, across a man-made lagoon connected to the canal, Piëch had built the Autostadt, or Auto City, a combination museum, dealership, and temple of automotive technology, which opened in 2000. Reached from Wolfsburg by a pedestrian bridge spanning the canal, the Autostadt had a dining complex with several restaurants as well as several floors of exhibits on the history of Volkswagen and the auto industry. Outside, on exquisitely manicured grounds tended by workers in identical brown coveralls, were pavilions celebrating all the Volkswagen brands, from Lamborghini to Skoda. Amid the rolling lawns were small ponds, some occupied by beavers, which had lost their fear of humans and sometimes lazed on the grass, helping underscore the company's assertion that it was a car company, yes, but also a friend of nature.

The Autostadt functioned as a kind of dealership where Volkswagen customers could take delivery of their vehicles. The finished cars waited inside two twenty-story cylindrical towers with transparent façades. Mechanical arms fetched the vehicles from bays, then lowered them carefully to ground level, like giant vending machines. The Autostadt even had its own hotel, a Ritz-Carlton, certainly one of the few five-star hotels in the world to overlook a car factory. The hotel had a restaurant with three Michelin stars, and guests could bathe in an infinity pool bordering the lagoon beneath the tall smokestacks of the Volkswagenwerk's power generation plant. Farther to the east was a new stadium, the Volkswagen Arena, home of VfL Wolfsburg, a professional soccer team owned by Volkswagen.

The amenities helped relieve the melancholy of Wolfsburg, which had grown from a tiny village to a city with 120,000 residents but offered little in the way of culture or entertainment. The attractions were also intended to make it easier for Volkswagen to compete for top engineers with BMW and Daimler and the charms of their home cities, Munich and Stuttgart. Long gone were the days when Volkswagen could outsource most of its development work to Porsche. For projects like the EA 189, Volkswagen needed to attract the best people available.

By 2005, well over ten thousand people worked in the sprawling development complex where new models and engines were designed. Located in an area known as Sandkamp, the complex had its own exit on the autobahn and its own oval test track. There was little architectural unity to the jumble of functional buildings that had been added over the years, except that they usually made use of the same dark brick as the main factory. The size of the R&D complex reflected how much more complicated it had become to manufacture cars. The need to save fuel had increased the importance of aerodynamics, while safety regulations required air bags and front ends designed to crumple on impact with minimum injury to drivers or pedestrians. Ever tighter controls on emissions in both Europe and the United States meant that tailpipe exhaust had to be purified as much as possible by the use of catalytic converters and particle filters. It was a constant struggle for automakers to strike a balance between fuel economy, customers' desire for performance and styling, and the demands of regulators for cleaner emissions and safer vehicles.

Like banks or any other highly regulated industry, carmakers needed strong compliance departments to keep engineers abreast of regulatory requirements and to ensure that they did not risk the legal and financial repercussions of cheating. Pischetsrieder, according to people who worked with him, recognized that Volkswagen was behind BMW and Daimler in its internal controls. He tried to use the prostitution scandal as an opportunity to introduce reforms. But before he could complete the work, he fell out of favor with Piëch, who appeared to feel threatened by Pischetsrieder's increasing independence and his efforts to remake Volkswagen culture along less authoritarian lines.

Pischetsrieder was popular with shareholders because he had

improved Volkswagen's profitability and begun to address the efficiency gap with Toyota. Volkswagen's share price had nearly doubled under his leadership. But there was no way to cut costs without disturbing the empire assembled by Piëch, who was still chairman of the supervisory board. Pischetsrieder had already dared to tinker with one of Piëch's pet projects when he canceled sales of the Phaeton in the United States because of poor demand. Along with Wolfgang Bernhard, a hard-charging former Daimler executive in his early forties, Pischetsrieder also threatened the turf of the labor leaders. He tried to cut the German workforce by twenty thousand by means of an early retirement scheme and by selling subsidiaries. Bernhard even had the gall to try to close one of Volkswagen's unproductive factories—a direct affront to workers, even though the factory was in Brussels. The closure would not have directly affected jobs in Germany, but it might have set an unwelcome precedent.

When employee representatives began to turn against Pischetsrieder, whose five-year contract expired in 2007, Piëch did little to defend his handpicked successor. In an interview in March 2006, Piëch told the *Wall Street Journal* that renewal of Pischetsrieder's contract was an "open issue" because of worker opposition to his policies. That icy assessment, rather than an expression of support, signaled Pischetsrieder's doom. By the end of the year, he was gone, followed soon afterward by Bernhard.

The new chief executive appointed by the supervisory board was Martin Winterkorn, the man who had overseen development of the Phaeton and was head of Audi. In contrast to Pischetsrieder, Winterkorn had a reputation for being obedient to Piëch. He could be counted on to forcefully execute the chairman's wishes.

So in the end it was Pischetsrieder, the reformer, who wound up leaving in the wake of the prostitution affair. Piëch emerged, if anything, more powerful than ever. With worker support, he had rid himself of a chief executive who had challenged his legacy, and installed a loyalist instead.

But with Pischetsrieder's departure Volkswagen also lost an executive who might have strengthened internal controls, set a higher standard of ethical behavior, and created an atmosphere where employees could be more open about their views. Had Pischetrieder stayed, the company's history might have turned out differently.

Volkswagen did follow through on Pischetsrieder's plan to hire ombudsmen to whom employees could report problems. But, as later events would prove, internal controls remained too weak to prevent wrongdoing, including misconduct with the potential to destroy the company. What mattered most was the example set at the top.

CHAPTER 10

The Cheat

SALACIOUS AS IT WAS, the prostitution scandal had no visible effect on Volkswagen's ambition. Quite the contrary, in November 2007, even as the affair was still playing out in court, the magazine *Der Spiegel* published a short item about an upcoming Volkswagen supervisory board meeting that included a startling assertion. Martin Winterkorn, not yet a year in office as chief executive, planned to present the board with a plan to push sales of Volkswagen cars and trucks to more than ten million in a decade. At the time, Volkswagen was selling about six million cars a year. Anyone who knew the auto industry and did the math would see the implications. To meet Winterkorn's sales target, Volkswagen would have to blow past number two automaker General Motors and then number one Toyota and become the biggest carmaker in the world. It was official. Volkswagen was bent on world domination.

Diesel was central to the plan. Volkswagen had already begun a major push to sell small cars powered by diesel engines. A few months earlier, at the International Motor Show in Frankfurt, Volkswagen had unveiled six new diesel models, including a Jetta and three new versions of the Golf. The cars were clean, practical, and economical, said Volkswagen, which boasted about how much less carbon dioxide the motors produced.

The United States was also a big part of the plan. It was the largest car market on the planet. To sell ten million cars a year and overtake GM and Toyota, Volkswagen could not ignore the United States. Beyond the business logic, the United States was an embarrassment for Volkswagen. Volkswagen was the dominant carmaker in Europe. It was a powerhouse in China and Latin America. But in the United States it was little more than a niche brand, in the same league as imports like Subaru.

Behind the scenes, there were problems. It had been a tumultuous year as Volkswagen struggled to design and produce the EA 189, the new diesel motor that was crucial to Volkswagen's ambitions. Before his ouster as Volkswagen chief executive the year before, Bernd Pischetsrieder had planned to equip Volkswagen diesels with emissions technology borrowed from Daimler, the maker of Mercedes-Benz cars based in Stuttgart. Wolfgang Bernhard, a former Daimler executive who was head of the division that made Volkswagen brand cars, was also an advocate of using the technology. Known as BlueTec, it used a solution containing the chemical urea to break down nitrogen oxides in the exhaust into a harmless form of nitrogen and oxygen.

THE TECHNOLOGY WAS EFFECTIVE, but had a number of drawbacks. The cars had to be equipped with a tank to hold the urea solution, which owners or their mechanics would need to refill periodically. An extra tank might not seem like a big deal, but from Volkswagen's point of view the effect on sales—and company ambitions—was significant. The urea tank would take space from the cargo area. That in turn might affect the ratings that the car got from auto magazines, whose reviewers are zealous measurers of cargo capacity. In addition, refilling the tank would become an extra chore and expense for the owner, a potential turnoff for prospective customers. BlueTec also added as much as $350 to the price (some estimates are lower), which could be a huge competitive disadvantage in the crowded market for midsize sedans. All these factors were especially acute in Volkswagen's case, because it planned to put diesel at the forefront of its drive to build market share in the United States. Volkswagen could not afford anything that might take away from diesel's appeal.

Pride also played a role. It was painful for the engineers in Wolfsburg and at Audi in Ingolstadt to borrow technology from Daimler (which was then still known as DaimlerChrysler). They saw themselves as pioneers in diesel. It was, after all, Audi that in 1989 had brought the first turbo-charged, direct fuel injected diesel engine to the mass market.

Some engineers at the Wolfsburg development complex sincerely believed that the Daimler technology was not yet mature enough for the mass market. There were genuine shortcomings. For example, the urea emissions cleansing system was not effective until the engine warmed up. And what worked for Daimler didn't necessarily work for Volkswagen. Mercedes were luxury cars, where the extra cost of the technology made less of a difference. They were bigger, too, with more room for the urea tank.

Ferdinand Piëch, still chairman of the supervisory board, may also have been suspicious of the motives behind the cooperation with Volkswagen's Stuttgart rival. For years, there had been rumors that Daimler was interested in taking over Volkswagen, or building up a big enough stake to exert influence. Christian Wulff, the prime minister of Lower Saxony, who had a tense relationship with Piëch, had openly encouraged Daimler to buy Volkswagen shares. Piëch could easily have concluded that there existed a conspiracy to open the gates in Wolfsburg to Daimler and undercut his power, and that the BlueTec alliance was part of it.

The only practical alternative to BlueTec was called a lean NOx trap, or LNT. A form of catalytic converter, it trapped nitrogen oxides in a chamber and separated the molecules into oxygen and diatomic nitrogen, the harmless form of nitrogen found abundantly in the earth's atmosphere. Because the chamber quickly became saturated, it was periodically regenerated by briefly increasing the ratio of fuel to air inside the cylinders where ignition takes place. Surplus unburned fuel coming from the engine into the NOx trap acted as a kind of cleaning agent, flushing out the trapped nitrogen.

The lean NOx trap technology was less expensive than Daimler's chemical solution and, because there was no urea tank to refill, required no maintenance by owners. But, as is often the case with emissions technology, solving one problem created another. Because the lean NOx trap was not capable of completely neutralizing nitrogen oxide emissions, the

Volkswagen system also used a third pollution control technology, called exhaust gas recirculation, or EGR. As the name implies, the system recycled some of the exhaust gas, pumping it back into the cylinders. The exhaust gases, with less oxygen content than the air in the atmosphere, lowered the combustion temperature inside the cylinders, which in turn led to lower production of nitrogen oxides. (Engines produce more nitrogen oxide at higher temperatures. That's why diesel motors produce more of it than gasoline engines, which operate at lower temperatures.)

The drawback of exhaust gas recirculation was that it caused the engine to produce more cancer-causing fine particles. That in turn put a heavier burden on another component of the emissions control system, the filter that trapped the soot particles and prevented them from exiting the tailpipe. As tests in Volkswagen's in-house emissions lab revealed, the extra soot could cause the filter to wear out too soon. That was a problem both for regulators and for customers. By law, emissions systems are to remain effective for the useful life of the car, defined in the United States at the time as 120,000 miles. If the filter had to be replaced after, say, 50,000 miles, it might not be in compliance with regulations and would also create an extra expense and inconvenience for owners, potentially hurting sales. These problems were more difficult to solve in cars sold in the United States, where standards on nitrogen oxides emissions were stricter than in Europe, and where the penalties for malfeasance were far more severe.

Volkswagen's sorry position in the United States in 2007 stemmed from years of missteps, often the result of refusal by management in Wolfsburg to cater to American tastes. The Beetle's success decades earlier had been largely an accident. Ferdinand Porsche designed the original Volkswagen with no thought at all for the American market. The only customer who mattered was Adolf Hitler. More than half a century later, though, managers and executives in Wolfsburg remained surprisingly ignorant of the American market. They were focused on Europe and were notorious for refusing to listen to advice from their people abroad.

Walter Groth, who worked for Volkswagen in the United States in the 1990s, tells a story about the lengths he went to in order to get developers in Wolfsburg to install cup holders big enough for American takeout cups. In Germany, a cup of coffee is seen as something to be enjoyed in civilized fashion, sitting at a table. At least until the arrival of Starbucks in the 2000s, the whole idea of takeout coffee was, to German eyes, uncouth. Despite pleading from American dealers, Volkswagen executives refused to believe that Americans eat entire meals while driving, and to design car interiors accordingly. "It was impossible to explain in Wolfsburg that our cup holders didn't work," Groth recalled.

So one morning Groth organized a convoy of Volkswagens for a delegation from Wolfsburg that was visiting Los Angeles. He took the visitors to the drive-in window of a McDonald's, had them order coffee and breakfast, then insisted that they eat their Egg McMuffins at the wheel. The lesson penetrated. Eventually Volkswagen made cars with cup holders ample enough for American fast-food portions.

Volkswagen needed a way to set itself apart from Toyota in the States. From a Wolfsburg perspective, the choice was obvious: diesel. The Volkswagen TDI offered fuel economy almost as good as that of the Toyota Prius, and no one else in the U.S. market had it. Diesel had been Volkswagen's springboard to dominance in Europe and could play the same role again. The decision to commission the new EA 189 motor in the first place was largely driven by Volkswagen's ambitions in America. Existing motors had no chance of meeting stricter U.S. regulatory limits on nitrogen oxide emissions.

But production of the new motor was delayed by an intense, internecine debate about what kind of fuel injection system the new motors should employ. Many people inside Volkswagen did not want to abandon the existing technology, known as "unit injector," or *Pumpe Düse* in German. The system had been groundbreaking for its time, allowing fuel to be pumped into the cylinders at 2,000 bar, or 2,000 times the pressure of earth's atmosphere. That was nearly twice as much pressure as previous technologies used, and it allowed the fuel to be burned more efficiently and with fewer emissions. Ferdinand Piëch had played a personal role in the development of *Pumpe Düse*.

But common rail was becoming the industry standard because of its superior efficiency. With *Pumpe Düse*, after each squirt of fuel, the injector nozzle had to shut off to allow fuel pressure to regenerate. With common rail, pressurized fuel was continuously available, permitting the timing of the injection to be calibrated more precisely. The result was better fuel economy and lower emissions. Common rail also cost less. The rest of the auto industry was moving to common rail. The increased demand lowered the price, because suppliers of common rail systems like Robert Bosch could produce it in greater numbers and spread development and tooling costs among a greater number of customers.

Despite those arguments, it was excruciating for Volkswagen to give up its proprietary technology. One engineer, who worked in Wolfsburg at the time and insisted on anonymity, described the internal debate as "a holy war." Piëch was known to dislike copying the competition, and his acolytes were among the *Pumpe Düse* holdouts. So sacred was *Pumpe Düse* inside Volkswagen that the engineers who favored common rail did the initial development in secret, working in a basement at the sprawling development complex, because of fear of the consequences if top management found out. The argument was settled after an internal study showed that common rail would save well over €1 billion a year, or about $1.3 billion, in procurement costs.

By then, the new engine project had fallen two years behind schedule. Even after pushing back the launch date, less than two years remained before the scheduled start of production of the new motor. That is an exceptionally tight time frame in which to design a completely new motor, set up an assembly line to make it, while also solving the emissions problem. Cars with the new U.S. motor, a variant of the EA 189, were supposed to begin arriving on the lots of American dealers in late 2008.

Meanwhile, Winterkorn's promotion to chief executive in 2007 brought an end to a period of perestroika in the way Volkswagen was managed. Pischetsrieder had tried to loosen the rigid, top-down style of management he had inherited from Piëch. Winterkorn, who owed his advancement to Piëch, reinstated it.

Winterkorn differed in many ways from his mentor. Whereas Piëch came from privilege, Winterkorn came from humble circumstances. His parents were ethnic Germans who had lived in Hungary and fled after the end of World War II to Germany, where he was born in 1947. Winterkorn studied metallurgy and physics at the University of Stuttgart, later earning a doctorate in metal physics at one of the prestigious Max Planck Institutes. He began his working life at Bosch, and joined Volkswagen in 1981 as a quality controller at Audi. He worked alongside Piëch for thirty years, once describing their relationship as a partnership where Piëch was responsible for conceiving new ideas while Winterkorn was in charge of executing them. Despite decades of working together, the two maintained a professional distance. They addressed each other with the formal German word for "you," *Sie*, rather than the informal *Du*.

They had one thing in common. Both ruled by fear. Piëch, relatively slight of build and slender, could demolish someone who displeased him with a few cutting words or simply a cold stare. There was nothing subtle about Winterkorn. A former soccer goalkeeper, he was physically intimidating, with a stocky build and a way of walking that thrust his chest forward. He wore double-breasted suits with wide lapels that emphasized his bulk. When Winterkorn was unhappy about something, which was often, he conveyed his feelings loudly and sometimes physically. At one meeting to discuss components that would go into the U.S. version of the Passat, Winterkorn became so angry about a part that he considered second-rate that he banged it on a table until it broke, according to an executive who was present. Several other people who worked with him described similar scenes. Yet Winterkorn was also someone who seemed to crave respect. He liked to be referred to as "Professor" Winterkorn, although his professorships were all honorary.

To be sure, some of the more strong-willed executives were not intimidated by Winterkorn's bluster. Many forgave his outbursts, rationalizing them as an expression of his commitment to quality. Winterkorn's bark was worse than his bite, they said, and he could be personable and open with people he trusted. But others feared Winterkorn's wrath and were not willing to risk the consequences of telling him that something he wanted was not attainable. And everyone knew that Winterkorn had a close relationship with Piëch, who was still chairman of the supervisory board.

This was the environment that prevailed in late 2006 as engineers, executives, and software experts involved in development of the U.S. variant of the new diesel motor struggled with the emissions problem. Pischetsrieder was on his way out, and Bernhard soon followed him out the door, eventually returning to Daimler. Winterkorn, with Piëch in the background, was asserting his power. Volkswagen was preparing to make a huge bet on the United States, the keystone of its plan to become the largest carmaker in the world. BlueTec was out of favor, if not yet officially dead.

In mid-2006, the engine developers realized they had a big problem. Tests in Volkswagen's own labs revealed that the exhaust gas recirculation system in the new motor would place an unacceptable burden on the particle filter and cause it to wear out prematurely. Yet the lean NOx trap was not capable of keeping nitrogen oxide emissions under control without help from the EGR system. The cars would not be able to pass emissions tests, especially in the United States, with its stricter limits on nitrogen oxides. The Environmental Protection Agency would not certify the cars, which would therefore not be able to go on sale. Volkswagen's global ambitions had bumped up against the laws of physics. As Volkswagen later admitted in court documents, its specialists were not able to reconcile the conflicting goals of fuel economy and emissions "within the allocated time frame and budget." An engineer who was involved in development of the new motor put it more bluntly. "It was a bad plan," he said.

Volkswagen was not without choices. For example, it could have put a special warranty on the particle filter, entitling customers to free replacement of the filter when it wore out. That would have made it easier for the pollution control equipment to operate at capacity. Or Volkswagen could have built cars with better emissions technology. Faced with a similar quandary, BMW equipped diesel SUVs sold in the United States with more robust equipment adequate to meet U.S. pollution standards legally. Some BMWs had all three emissions technologies: exhaust gas recirculation, lean NOx traps, and the urea-based SCR systems. But for Volkswagen cars aimed at a less affluent customer, the extra gear would have added hundreds of dollars to the vehicles' costs and taken up valuable space without adding to the cars' appeal. Emissions technology is not something dealers talk about when they try to sell a car.

Volkswagen chose another route to solve the problem. In mid-2006, engineers in Wolfsburg were working on adapting software used by Audi in its diesel motors for Volkswagen engines. Audi already had motors that used common rail systems, so it made sense to borrow the software rather than start from scratch. As the engineers explored the software, which contained thousands of functions used to control engine parameters, they noticed something labeled in English as a "noise function," also referred to as an "acoustic function." It allowed a car to recognize when it was being tested in a lab on rollers. If the computer under the hood sensed that the car was being observed, it could adjust the engine's behavior to deliver optimal test results. The software, the engineers realized, was a defeat device.

Someone—it's not clear who—mentioned that Volkswagen could use the function to solve its emissions problem with the new diesel motor. The comment might have been meant only half seriously, but it gained traction. In November 2006, according to class-action

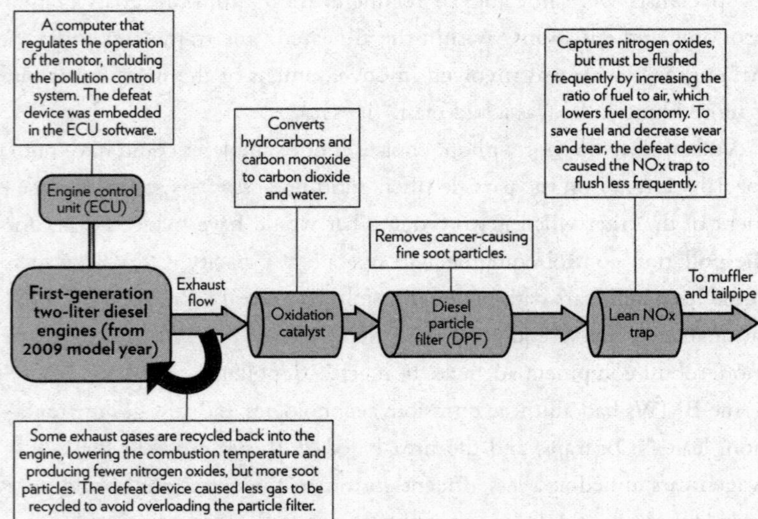

complaints later filed by Volkswagen owners, which in turn were based on internal company documents, an engineer who was responsible for the software that would be used in the U.S. diesel motors was asked by his superiors to formulate a way for the Audi acoustic function to be used in Volkswagens. The engineer was profoundly uneasy about the assignment. But he did as he was told.

That same month about fifteen people responsible for diesel engine development and engine electronics gathered in a conference room on the top floor of the Research and Development building in the Wolfsburg factory complex. The seven-story building is a long rectangle with a façade of red brick and white paneling. Hundreds of people work there, and the complex has its own cafeteria known for its curry wurst, a German specialty consisting of slices of sausage smothered in ketchup and curry powder.

The conference room, close to offices of the top R&D executives, was well appointed, with a plush carpet and polished wooden table. The developers took their seats. The highest-ranking executive present, according to information later gathered by investigators, was Rudolf Krebs, the head of motor development. The PowerPoint presentation that the software engineer had reluctantly prepared was beamed onto a screen. The presentation was just three pages long. (Tediously long PowerPoints were taboo in R&D.) One slide consisted of a graphic depicting the procedure used by the Environmental Protection Agency to test the amount of pollution spewing from a car. In a laboratory, with a test vehicle running on rollers, regulators would try to replicate the conditions a car would encounter on the road, including simulations of city and highway driving. The pattern of this so-called test cycle was public knowledge and predictable. The presentation explained how code embedded in the engine-control software could recognize the pattern, activating equipment to reduce emissions whenever a test was underway. The rest of the time, the pollution controls could be dialed back to protect the particle filter. In other words, the car would behave lawfully only when the emissions police were watching. The program described in the PowerPoint fit the definition of a defeat device, and it was illegal in both the United States and the European Union.

The people in the conference room were aware that they were taking a big risk. There was an intense debate about whether to deploy the software. Some people had serious reservations. They worried that the software could be detected by U.S. regulators and expose Volkswagen to legal problems. Some of those present felt that the cheating was simply wrong. Among them were idealists who truly believed they were working to build a cleaner engine. The idea of cheating was demoralizing; it was not what they had signed up for. Others argued that all the carmakers cheated. Volkswagen had to take shortcuts, too, or it wouldn't be able to compete, they said. At the end of the meeting, Krebs announced the decision, according to documents that later came to light. Build the defeat device. As the meeting broke up, he admonished the engineers not to get caught.

Krebs, who moved to a different position in 2007, later denied that he knowingly approved software to manipulate emissions readings at this meeting or at any other time. He maintains that the managers responsible for the development of diesel motors assured him that they could develop engines adequate for U.S. emissions standards, and that he first heard of the existence of a defeat device years later, in September 2015. A lawsuit later filed against Volkswagen by lawyers with access to internal company documents actually described Krebs as being skeptical about the use of software to pass emissions tests, because of potential legal problems it could cause.

The meeting lasted less than an hour. The consequences would be felt for years.

Probably none of the participants foresaw how dire the consequences would turn out to be. In Europe, the risk was not that high. There were almost no penalties for carmakers that violated the emissions rules. The United States, where penalties were routinely in the tens or hundreds of millions of dollars, presented a different story. But, ensconced in Wolfsburg, the engineers believed there was little chance of getting caught. As far as they knew, there was no technology available to measure emissions from a car that was driving on the road, exposing the huge discrepancy between the amount of pollution that the car produced in the lab and what it produced under normal driving conditions.

The people discussing the PowerPoint presentation may have seen the software as a temporary stopgap, a way to keep the engine program moving until a better solution could be found. As far as is known, none received any extra money for solving the emissions problem. The reward was that they got to keep their jobs.

Nor was there likely any sense among the developers that their actions were a grave violation of Volkswagen standards. There was plenty of precedent for using shortcuts to cope with inconvenient regulations. Volkswagen had a history of run-ins with U.S. environmental officials, but none had led to serious fines and they could be written off as the cost of doing business. Volkswagen had been accused of deploying defeat devices as far back as 1973. The company at that time paid $120,000 to settle with the Environmental Protection Agency. In 2005, Volkswagen paid a $1.1 million penalty for failing to notify the EPA of emissions problems in some cars manufactured in Mexico.

Audi, a breeding ground for many of Volkswagen's most important innovations, had been using software to evade emissions rules for years. Beginning in 1999, Audi engineers had invented a technology known as pilot injection, which eliminated the disagreeable clattering noise that diesel engines made when they started up. But the noise reduction technology caused emissions to increase above legal limits. So Audi developed software that recognized the simulated driving cycle used for official emissions tests in Europe. When the car detected a test, the noise reduction technology was deactivated. The rest of the time the car was programmed to spare customer eardrums. Audi called the software euphemistically the "acoustic function." Audi installed it in three-liter V-6 diesels sold in Europe from 2004 to 2008 to get around increasingly more stringent pollution limits. That time period coincided with part of Winterkorn's tenure. He was chief executive of Audi from 2002 to 2006. (Given Winterkorn's attention to detail, it is hard to believe he would not have been aware of the acoustic function. But no documents have come to light proving that he was aware of the noise-suppressing software.)

The Volkswagen executives and engineers simply adapted the acoustic function for their new diesel engine. The code itself was inside the so-called engine control unit, or ECU, which sat under the hood and managed the complex interplay of components inside the motor. Bosch, based in Stuttgart, had just developed a brand new ECU it called the EDC17. "EDC" stood for "electronic diesel control." Bosch advertised that the new computer could "be used very flexibly in any vehicle segment on all the world's markets." The EDC17 "offers a large number of options such as the control of particulate filters or systems for reducing nitrogen oxides." By tailoring the timing of fuel injection and other parameters, the EDC17 "improves the precision of injection throughout the vehicle's entire service life. The system therefore makes an important contribution to observing future exhaust gas emission limits," Bosch said.

After the meeting, lower ranking engineers prepared specifications for the defeat device which were given to Bosch. Volkswagen did not write its own software at the time, leaving the task of coding to subcontractors. Bosch's new EDC 17, developed largely with Volkswagen in mind, provided a convenient platform.

Bosch worked closely with Volkswagen to tailor the ECU for the EA 189 motor, while maintaining tight control over the code. It would have been difficult, if not impossible, for Volkswagen to modify the software without Bosch's approval. Bosch betrayed concern that it was aware of the legal risks. In June 2008, a Bosch executive wrote to Volkswagen warning that use of a defeat device was prohibited by U.S. law, and asked Volkswagen to indemnify it against any legal consequences. Changes to the software requested by Volkswagen would mean "yet another path toward potential input of data as a 'defeat device,'" the Bosch letter said. It went on to quote U.S. and European laws prohibiting defeat devices. "We request that your firm sign the enclosed indemnity clause," the letter said. As far as is known, Volkswagen refused. Instead, a Volkswagen executive in engine development scolded Bosch for getting lawyers involved. Bosch later argued that the letter referred to a different group of engines than the diesels with defeat devices.

Written or unwritten rules of behavior that would have restrained the Volkswagen engineers were weak or nonexistent. As Volkswagen later admitted, there was a tolerance for breaking the rules and a lack of checks and balances. The people who wrote engine software were the same ones who approved it for use; other auto companies, in contrast, separated software development and approval. Volkswagen's legal department did not have people with the technical knowledge that would have made clear what the engineers were up to. Volkswagen's legal and compliance staff also apparently did not realize the magnitude of the penalties that the company would face in the United States if the cheating was discovered, as opposed to the relatively light consequences it would have faced in Europe, so they could not impress that danger on the engineers. Volkswagen did not yet have a system in place that would allow an internal whistle-blower to report the violation without fear of consequences. Employees who had reservations about the illegal software—and there were some—had no place to turn. And top executives were hardly setting an example of moral probity. In the background, the prostitution scandal was still making headlines.

There were voices inside the company warning of the risks. Volkswagen engineers and executives had already been discussing the new diesel engines with EPA and CARB, the influential agency that enforces California's especially strict air pollution rules. The U.S. officials had reminded the Volkswagen people of their duty to be open about any circumstances when emissions control equipment might be partly or wholly disabled.

In November 2006, Stuart Johnson, a Volkswagen executive in the United States with responsibility for emissions compliance, wrote a memo that called attention to the case in the late 1990s in which Cummins, Caterpillar and other truck engine makers had been caught using defeat devices. Johnson recalled that the truck companies had programmed engines to pollute more after a certain amount of time had elapsed, which happened to be just a little longer than the federal testing procedure lasted. What the truck companies had done was a "clear violation of the spirit of the emission regulations and the certification test procedure," Johnson wrote. A few days later, Leonard Kata, another Volkswagen manager responsible for emissions regulations in America, reinforced Johnson's message in an e-mail. Anything that caused the car to behave differently under normal conditions than it did in a testing facility would be illegal, Kata said in essence.

On March 21, 2007, CARB officials repeated the message during a meeting with a delegation from Volkswagen that included high-ranking managers from Volkswagen and Audi in Germany. According to Volkswagen's internal account of the meeting, Duc Nguyen, a CARB official, told the Volkswagen people that the agency "expects emission control systems to work during conditions outside of the emissions tests." Volkswagen's report of the meeting added, "Volkswagen agrees."

Among themselves, many engine developers resented air quality regulations that, in their view, were becoming impossible to fulfill. The car companies faced pressure to lower carbon dioxide emissions to help slow global warming. Fuel-efficient diesel, in the view of the German engineers, was a promising way to fight climate change. Yet the regulations also required the carmakers to keep reducing diesel's most troublesome byproduct, nitrogen oxides. The two goals, less carbon dioxide and lower nitrogen oxide emissions, conflicted with each other, and both were hard to reconcile with the demands of the market. Customers wanted cars that were fun to drive, and few cared about what kind of emissions system they had.

In 2007, Wolfgang Hatz, the head of engine development at Volkswagen and a confidant of Winterkorn's, gave voice to those frustrations during a technology demonstration in San Francisco. Hatz was particularly critical of the state of California and its very strict air quality standards.

"CARB is not realistic. It is too aggressive," Hatz said in remarks captured on video by an automotive website. The sharp reductions in emissions demanded by CARB are "nearly impossible for us," Hatz complained. The video shows him wearing a pink dress shirt open at the collar, with a white undershirt visible underneath and a bit of stubble on his face. "We have to be realistic," Hatz said. "From our company, from our industry we will do what's possible. We can do quite a bit and we will do a bit, but impossible we cannot do." Hatz painted a car guy's vision of dystopia: a future when Americans had no choice but to drive tiny economy cars. "Perhaps we have just small Japanese and Korean cars in this country," he said, making it sound like a threat.

In August 2007, Audi and Volkswagen officially killed the alliance with DaimlerChrysler on emissions technology. The decision meant that the first generation of Volkswagen diesels with the new EA 189 motor would have to pass emissions tests without the urea-based system. "The TDI brand is strong enough in the U.S.," an Audi spokesman told the trade magazine *Automobilwoche*. "We don't need BlueTec." Audi did not, of course, explain what technological breakthrough had allowed the spokesman to be so confident.

Unlike the Volkswagen brand cars, some of the larger Audis would still get SCR systems. But Audi undermined the systems' capabilities by setting requirements that put the division's engineers in a bind. In 2006, Audi engineers realized that Q7 SUVs would need bigger urea tanks to meet U.S. limits on nitrogen oxides. Or owners could be required to bring their vehicles to the dealer for urea refills every five thousand miles or so. Both alternatives were rejected. Instead, the Q7s were also programmed to recognize the telltale pattern of an official testing procedure. When a test was underway, the Audis sprayed more urea into the pollution control system so that nitrogen oxide emissions would be within legal limits. When no test was detected, the engine software rationed urea fluid so that it would last at least ten thousand miles. Emissions rose accordingly, to as much as nine times the legal limit.

Winterkorn, still in charge of Audi at the time, was informed in 2006 that urea tanks in the Q7 were too small. So was a certain H. Müller, possibly short for "Herr Müller." Matthias Müller, who was to become a very important person in Volkswagen history, was then head of product management at Audi (although he denies any knowledge of the illegal software prior to September 2015).

No evidence is currently available proving that Winterkorn ordered use of a defeat device in the Q7 or was explicitly aware of one at that time. But documents indicate that the decision not to equip Q7s in the United States with bigger urea tanks, or to require more frequent refills, was made at the highest levels. The decision left Audi engineers with little choice. They could not do what was physically impossible. They could not configure the Q7 to meet U.S. emissions standards without draining the urea tank well before the car was due for an oil change.

They could admit failure to their superiors, or they could cheat. Faced with possible loss of their jobs, they chose the second option.

The Q7 diesel went on sale in the United States in late 2008. Porsche Cayennes and Volkswagen Touaregs, which had the same three-liter engine and emissions technology, later deployed the same defeat device. It was effective enough to fool U.S. regulators for years to come.

Even with the emissions hack, Volkswagen engineers in Wolfsburg struggled to complete the EA 189 program in time for Jettas with the new engine to go on sale in late 2008. But by April 2008 the motor was far enough along that Volkswagen could unveil it at the Vienna Motor Symposium, an annual showcase for new technology that regularly attracts Europe's top auto executives. The congress was held at the Hofburg, a Baroque palace in Vienna that was once the seat of the Habsburg Empire.

Volkswagen was proud of what it had achieved, or appeared to have achieved, with the new EA 189. The presentation to congress participants was entitled "Volkswagen's new 2.0 liter TDI engine fulfills the most stringent emissions standards." As the symposium drew to a close, Winterkorn gave a speech that took the boast further. "The steep pace of mobility growth presents enormous challenges to the environment and infrastructure," Winterkorn said from behind a podium, dressed in a crisply tailored dark suit, a slender wireless microphone clipped to his ear. "The Volkswagen Group does not shirk from its responsibility in this context." The company, he went on, was "searching worldwide for innovative solutions to bring down emission levels through ever more efficient internal combustion engines and gearboxes and through the development of alternative fuels and powertrain concepts." He promised that Volkswagen would "set new benchmarks for high fuel economy and environmentally sound motoring." The company, Winterkorn concluded, "is pursuing a global objective: to reconcile sustainability and mobility around the globe."

The speech described the contours of how Volkswagen wanted to be seen by the world. It was a grand new vision, from a man known more for his obsession with engineering details than for an expansive worldview. Volkswagen would be not only the biggest car company on the planet but also the greenest.

CHAPTER 11

The Porsches and the Piëchs

THE PORSCHE SPORTS CARS and SUVs manufactured with Volkswagen made the Porsches and the Piëchs into some of the wealthiest families in Europe. Still, there was chronic discord at the rustic mountain estate in Zell am See known as the Schüttgut, the Porsche and Piëch family base. While Louise Piëch and Ferry Porsche, who died in the late 1990s, had a good working relationship, their children divided into bickering camps. The Piëchs, raised under the stern guidance of Louise, were known for being tough and ruthless. They attended strict Swiss boarding schools. Ferdinand Piëch was this branch of the family's most prominent scion by far. Ferry Porsche, Louise's brother, had a reputation for being more laid-back and avuncular. His children attended Waldorf schools, which had been founded by the esoteric Austrian philosopher Rudolf Steiner and strove to offer children a nurturing, egalitarian environment. The Waldorf schools were a reaction to the hierarchical school system in German-speaking Europe, which separates students at an early age into those bound for universities and those bound for blue-collar jobs.

Stereotypes about the two branches of the family—the steely Piëchs and the more sensitive Porsches—were somewhat exaggerated. Ferry, after all, had spent World War II at his father's side designing tanks for Hitler. And subsequent events would show that the Porsches could also be Machiavellian. But later generations of the Porsches and Piëchs were different enough that they regarded themselves as separate families, even if the outside world saw them as a unit. Reporters were scolded by family public relations staff not to refer to the Porsche family, but rather to the Porsche *and* the Piëch families. Wolfgang Porsche, one of Ferry's sons, became the leader and spokesman for the Porsches, while Ferdinand Piëch was the dominant voice among the Piëchs. In the rivalry between Ferdinand Piëch and Wolfgang Porsche, Piëch—with his infighting skills and superior knowledge of automobiles—usually came out ahead. One thing that the two sides had in common was their secretiveness, for they almost never gave interviews and avoided public behavior that would provide fodder for the gossip columns. When representatives of the families met to resolve business issues, it was often in second-rate hotels or restaurants where no one would expect to find them. Mobile phones were banned from the meeting room.

Despite the friction, the family members were capable of pulling together when circumstances demanded it. Such an occasion came in the mid-2000s at the same time that the prostitution scandal was rattling Volkswagen and the company was preparing its renewed assault on the United States.

Volkswagen shares had languished in the final years of Piëch's tenure as chief executive. With its meager profit margin, Volkswagen offered little excitement to investors, despite its position as the leading automaker in Europe. Piëch was more interested in cars than in money; he preferred to invest profits in research and development rather than to pay dividends. He also refused to genuflect to financial analysts and the business press. Now his disdain for the stock market made Volkswagen vulnerable. In 2004, Volkswagen preferred shares (VW also had both nonvoting preferred and voting ordinary shares) fell below twenty-two euros, a 50 percent decline since 1998. The shares closed 2004 down 15 percent compared with the end of 2003, even as auto stocks in general recorded gains for the year. In its annual report, Volkswagen blamed the decline on what it said was an unfavorable environment for carmakers. But the company conceded that "investors were also disappointed by Volkswagen's earnings performance."

The company's market capitalization, the stock market value of the preferred and ordinary shares together, was just €13.3 billion ($18 billion) at the end of 2004. Volkswagen was, in other words, a bargain and ripe for a hostile takeover. The only thing protecting the company was the special law, the Volkswagen *Gesetz*, passed as part of the political compromise in 1960 that allowed the federal government to sell its shares on the stock market. The law limited the voting rights of any one shareholder to 20 percent, regardless of how many shares the investor might own. In addition, 20 percent was enough to veto decisions made at the annual meeting. The Volkswagen law made a hostile takeover all but impossible, since the state of Lower Saxony owned 20 percent and was not likely to give up control of the state's biggest employer. But the European Commission, the body that administers the European Union, saw the Volkswagen *Gesetz* as a violation of rules on the free movement of capital. The law created a privileged class of shareholders and discriminated against the others. In 2005, the commission sued to strike down the law.

For Porsche, the potential demise of the Volkswagen law was both a threat and an opportunity. The threat was that a new owner of Volkswagen might not be willing to continue the SUV partnership, at least not on such generous terms for Porsche. Without its exclusive access to Volkswagen factories, Porsche might become just another struggling niche sports car maker. Porsche was also threatened by ever stricter standards for carbon dioxide emissions and fuel economy. Both the European Union and the United States required carmakers to achieve fuel economy milestones based on fleet average—in other words, the average gasoline or diesel consumption of all the cars produced by a given company. With its fleet of powerful sports cars and SUVs, it would be extremely difficult for Porsche to keep pace with tightening standards. The problem would be solved if Porsche attached itself legally to Volkswagen. Porsche could continue to make sports cars and

SUVs if it could average the fuel consumption of its cars with Volkswagen's fleet of small, fuel-efficient cars.

The opportunity for Porsche in Volkswagen's low stock price lay in the financial firepower it had acquired because of its lopsided share of profits from the SUV partnership. If somebody was going to take over Volkswagen, why not Porsche? Still, there was widespread shock when, in September 2005, Porsche disclosed that it had acquired 20 percent of Volkswagen's voting shares, suddenly becoming Volkswagen's largest shareholder. "Porsche's strategic objective," the company explained, "is to assure the long-term security of its future plans insofar as they involve Volkswagen as a development partner, and at the same time to prevent financial investors from staging any form of hostile takeover of Volkswagen AG." Years later, when he had to explain the company's actions in court, Wiedeking was blunter. "For Porsche as a niche provider," he said, "it was practically a matter of survival."

THE IDEA THAT TINY PORSCHE could swallow a big chunk of Volkswagen was, from a financial perspective, not as outlandish as it seemed. With a market capitalization of €11.4 billion ($13.9 billion) in August 2005, Porsche was worth almost as much on the stock market as Volkswagen. Investors did not care that Porsche produced only about 100,000 cars per year, compared with more than 5 million for Volkswagen. What mattered to the stock market was that Porsche made more money—€1.4 billion ($1.7 billion) in its 2005–2006 fiscal year compared with €1.1 billion ($1.3 billion) for Volkswagen. The irony, of course, was that the profit margins that enabled Porsche to buy a big chunk of Volkswagen were largely the fruit of Porsche's SUV partnership with Volkswagen.

The question that would later occupy the German justice system was whether Wiedeking and his shareholders, the Porsche and Piëch families, had a secret master plan in 2005 to eventually take over Volkswagen, which they illegally concealed from investors. Wiedeking and members of the families adamantly disputed that accusation. It soon became clear, though, that Porsche did not intend to stop at 20 percent.

IN NOVEMBER 2005, Porsche bought more shares and raised its stake to 27 percent. The following March, Porsche raised its stake in Volkswagen to above 30 percent. That was the point when, according to German securities law, Porsche was obligated to make an offer for all outstanding Volkswagen shares. Porsche offered €100.92 per voting share and €65.54 for preferred shares, less than the market price at the time. Not surprisingly, few shareholders took the offer.

At the same time it was acquiring shares, Porsche also bought options designed to protect itself against future fluctuations in the price of Volkswagen shares. The options allowed Porsche to lock in the ability to acquire Volkswagen shares at a certain price within a given time period, regardless of how much Volkswagen's share price might fluctuate on the market. Porsche executed the options with help from the Frankfurt branch of Maple Bank, a small Canadian investment bank. The options proved to be a smart move—at least in the short term. As Volkswagen's share price climbed, Porsche was able to profit from the difference between the price it had locked in through options (which varied over time) and the market price.

The profits from options soon rose so steeply, in fact, that they began to eclipse profits from making cars. In Porsche's 2006–2007 fiscal year, which ran from August through July, the company reported €3.6 billion ($4.9 billion) in earnings just from options. It also collected a share of Volkswagen profits—€111 million ($150 million)—and recorded a paper gain in the value of its Volkswagen holding of €521 million ($703 million). After taxes, net profit for the fiscal year more than tripled from the previous year, to €4.2 billion ($5.7 billion), on sales of €7.4 billion ($10 billion)—a profit margin of nearly 60 percent. Wiedeking, with his 1 percent share of the profits, became immensely rich. But the outsize earnings from financial legerdemain also raised the question of whether Porsche had become a de facto hedge fund with a sideline in carmaking. Amid the exuberance that prevailed in financial markets at the time, nobody asked many questions about what kind of risks might come with all of this easy money.

Ferdinand Piëch's stance toward Porsche's growing influence over Volkswagen appeared to be a mystery even to his own relatives. As a family shareholder, he acquiesced in the initial acquisition of the Volkswagen shares. But he also appeared to be wary of, if not openly hostile to, Porsche's move on Volkswagen, which posed a threat to his own power. The size of Porsche's stake entitled it to two seats on Volkswagen's supervisory board, which were taken by Wiedeking and Holger Härter, Porsche's chief financial officer. The supervisory board, made up largely of worker representatives and local politicians, had a reputation for rubber-stamping Piëch's plans. Wiedeking's presence changed that. Possibly for the first time, there would be someone on the supervisory board whose knowledge of the automobile industry could rival Piëch's. Wiedeking was also outspoken. He had both the authority and the mettle to stand up to Piëch.

A power struggle was preordained, and it was not long before the two men began to argue bitterly during management meetings. Wiedeking spoke dismissively of the luxury brands that Piëch had acquired, calling Bugatti "an expensive hobby." Piëch, for his part, spread the word that he was growing disenchanted with the way Wiedeking was running Porsche. *Die Zeit*, an influential weekly, quoted Piëch as saying that he would not allow his lifework—meaning Volkswagen—to be destroyed by "an ordinary manager." It was an extraordinary putdown of Wiedeking, who, after all, had turned around Porsche and made the family members into billionaires. Tensions grew to the point where Porsche's acquisition of Volkswagen shares began to have the opposite of the intended effect. Instead of solidifying Porsche's relationship with Volkswagen, it disrupted it.

ON OCTOBER 23, 2007, the European Court of Justice struck down the Volkswagen law, ruling that the special veto rights enjoyed by the state of Lower Saxony were a deterrent to other investors and therefore amounted to an illegal restriction on the free movement of capital. The ruling, welcomed by Porsche because it opened up a path for it to dominate Volkswagen, only intensified the ill will between Piëch and Wiedeking. Two days later, they and members of the Porsche shareholders' committee met at Wolfgang Porsche's house in Munich to try to clear the air, but the animosity only grew worse. Wiedeking later described the atmosphere as "poisonous." By his own account, Wiedeking lost his temper and accused Piëch of deliberately undermining the cooperation between Porsche and Volkswagen for his own aims. The meeting broke up without resolving any of the disputes.

By the following February, though, the ever enigmatic Piëch appeared to have changed his mind about the Volkswagen acquisition. Wiedeking later maintained that he convinced the Porsche and the Piëch families that Porsche could not cement its influence over Volkswagen with anything less than 50 percent of the voting shares. On February 11, 2008, the family shareholders approved plans to eventually acquire another 20 percent of Volkswagen and therefore a majority. The vote, according to Wiedeking, was unanimous. In July, Piëch also assented to a decision by the family to give Porsche management conditional permission to raise the stake to 75 percent.

Yet, behind the scenes, Piëch continued to keep everyone guessing about whose side he was on. In 2008, employee representatives on Volkswagen's supervisory board, who feared that a Porsche takeover would dilute their influence, proposed creation of a subcommittee that would approve all contractual agreements between Volkswagen and Porsche. The subcommittee proposal was an obvious attempt to keep Porsche on a short leash, exploiting the company's dependence on Volkswagen factories. Piëch had the power under German law to block the move. As chairman of the supervisory board, he could cast a tiebreaking vote if, as expected, the ten worker representatives voted in favor of the oversight committee and the ten shareholder representatives voted against it. But when the supervisory board convened on September 9, 2008, to vote on the measure, Piëch unexpectedly failed to show up. He conveyed his vote by writing: an abstention.

The measure passed 10 votes to 9, with the workers voting in favor and the shareholders—minus Piëch—against. Outside the meeting room at the Wolfsburg factory where the supervisory board met, thousands of Volkswagen workers—called to action by their union—demonstrated against a Porsche takeover. Some carried signs attacking Wiedeking personally.

While these machinations were underway, Porsche continued to accumulate huge profits. At the end of the 2007–2008 fiscal year, Porsche reported net earnings of €6.4 billion ($8.6 billion) on sales of €7.5 billion ($10 billion). Pretax profit, at €8.6 billion ($11.6 billion), actually exceeded sales, an unheard-of occurrence. The reason was simple: Porsche was making more money from the options it bought on Volkswagen shares than from the sale of cars.

But it was already clear that conditions in financial markets were becoming much less favorable for that kind of financial engineering. In March 2008, the collapse of Bear, Stearns had provided a forewarning of the risks lurking in the market for subprime mortgages, a form of asset the investment bank had helped pioneer. Porsche pushed ahead with plans to acquire a majority in Volkswagen, but buying the additional shares became more difficult and took time. It was not until September 16, 2008, that Porsche acquired another 5 percent of Volkswagen's voting shares, bringing the total stake to 35 percent. That happened to be a day after the investment bank Lehman Brothers filed for bankruptcy, causing financial markets worldwide to seize up.

Despite the growing market turbulence, on September 22 a banker for Merrill Lynch presented Porsche executives with a plan that called for Merrill, the British bank Barclays, and the Royal Bank of Scotland to raise €20 billion ($28 billion) in credit to finance the purchase of more Volkswagen shares. The project was code-named Bavaria. By October 20, Porsche Automobil Holding SE, the holding company for the families' shares in the sports car maker, had built its stake in Volkswagen to almost 43 percent.

It is an axiom of investing that risk and reward stand in inverse proportion to each other. An investment that delivers huge profits automatically brings with it the risk of huge losses. As the financial crisis deepened, and Volkswagen shares were caught in the global stock market rout, the risks that Porsche had taken started to become clear.

In addition to buying shares of Volkswagen outright, Porsche had continued to use derivatives to lock up control of VW shares that it did not own. Porsche executed its derivatives investments with Maple Bank, which was so obscure that even many bankers in Frankfurt had never heard of it. Why Porsche depended so heavily on Maple Bank is a matter of speculation. But certainly confidentiality was one possible explanation. Under German law, Porsche was also not required to publicly disclose the options holdings. It is easier to keep a secret at a small organization than at a big one. Porsche was by far the biggest client of Maple Bank's Frankfurt branch, which occupied space in an unremarkable six-story brick-and-glass office building in the city's West End, a quiet, affluent neighborhood favored by boutique financial firms.

Porsche had an agreement with Maple Bank that now proved to be a liability. Porsche had purchased options from Maple, known as call options, allowing it to buy Volkswagen shares by a certain date at a certain price, known as the strike price. At the same time, Porsche sold so-called put options to Maple Bank, allowing the bank to buy Volkswagen shares from Porsche at a certain price by a certain date. In other words, Porsche had not only options to buy shares but also obligations to sell them. For every call option, there was a put option with the same strike price and time frame. In effect, the call options and the put options were mirror images of each other. For Porsche, the advantage of this structure was that it gave the company effective control over the Volkswagen shares without actually having to own them or pay the full price. In banking parlance, Porsche had "synthetic shares" in Volkswagen.

This strategy worked brilliantly when Volkswagen shares were going up. Porsche made money selling the put options. But the downside of the strategy became obvious in the wake of Lehman's collapse. Porsche's agreement with Maple Bank required it to post cash security with the bank if Volkswagen shares fell below certain benchmarks. The more Volkswagen shares fell, the more Porsche had to pay.

Initially, Volkswagen shares rose in the weeks after Lehman filed for bankruptcy, nearly doubling, to more than four hundred euros, by October 16, 2008. But then they began to follow other German shares downward. Hedge funds swooped in and started to bet that VW's price would fall further. The hedge funds, most based in the United States, used a risky short-selling strategy in which they sold borrowed shares in anticipation of buying them back later, at a lower price, and pocketing the difference. Compounding the risk, some sold shares that they hadn't actually borrowed—a "naked" short that makes it even tougher to cover the bet later.

As Volkswagen shares plummeted, Porsche's obligations to Maple Bank began piling up on an almost hourly basis. At 8:14 a.m. on October 21, for example, Maple Bank confirmed in an e-mail to executives in Porsche's finance department that it had received €300 million ($420 million) in security. At a few minutes after noon, a bank employee sent another e-mail to the Porsche executives advising them of their obligation to transfer another 243 million ($340 million). Porsche obliged forty minutes later. Then, at 2:10 p.m., Maple Bank asked for another €417 million ($584 million). And on it went.

The crisis came to a head on October 24, 2008. That was a Friday, the day when the calls and puts between Porsche and Maple expired and would be rolled over for another week. The strike price on options for 17 million shares was €362, reflecting Volkswagen's higher share price a week earlier. But the actual share price that day was closer to €210. In effect, Porsche would have to make up the difference between the higher and the lower share prices. The company's loss after the calls and puts were settled came to €2.5 billion ($3.5 billion). All told, according to figures later presented in court, the company had already had to post €4.3 billion ($6 billion) by October 24 and was down to its last €326 million ($456 million). If the figures were correct—and Porsche later insisted they were not—the company was on the verge of bankruptcy. Not only would the Porsche and Piëch families' attempt to build a stake in Volkswagen fall apart; they stood to lose their fortunes and their control over the sports car maker. If Volkswagen shares rose, on the other hand, Porsche would get the money back and the pressure would ease.

On October 26, a Sunday, the Porsche holding company issued a statement. The company disclosed that it owned 42.6 percent of Volkswagen voting shares outright and options equal to another 31.5 percent. In other words, Porsche had already locked up 74.1 percent of Volkswagen's voting shares and had set a goal of acquiring 75 percent, the amount needed to control the company under German law. Porsche maintained that it was disclosing the information out of consideration for the hedge funds betting against VW. "Porsche has decided to make this announcement after it became clear that there are by far more short positions in the market than expected," the news release said. "The disclosure should give so called short-sellers—meaning financial institutions which have betted or are still betting on a falling share price in Volkswagen—the opportunity to settle their relevant positions without rush and without facing major risks."

If Porsche had suddenly acquired a soft spot for hedge funds, it was a funny way to show it. Investors already knew that Porsche was aiming for a majority in Volkswagen and had acquired a sizable block of shares, as well as options. But the size of the options position came as a shock to the investors betting that the stock would fall. It didn't take long for the hedge fund traders to do the math. If Porsche had locked up more than 74 percent of VW's voting shares, and the state of Lower Saxony owned another 20 percent, less than 6 percent was still available on the market.

Now the hedge funds were the ones with a big problem. They suddenly needed to settle their short-sale bets, which required them to own Volkswagen shares. The Porsche statement left them with the impression that not enough were available to buy. It was a classic "short squeeze"—one of the largest in financial market history. On Tuesday, two days after the Porsche statement, Volkswagen's share price topped €1,000 ($1,400) as the panicked short sellers scrambled to buy at any price. For a few hours, Volkswagen passed Exxon Mobil as the world's most valuable company.

But was there really a shortage of Volkswagen shares? German prosecutors would later say that the market panic was unwarranted, and that Wiedeking and Härter deliberately misled the market in order to inflate Porsche's share price and avert financial disaster. Porsche was on the verge of running out of cash it needed to sustain the options play, according to the prosecutors, and soon would have been forced to dump millions of shares on the market. In the October 26 statement, Porsche did not disclose the details of its agreement with Maple Bank or the existence of the put options, information that would have given the short sellers an inkling of Porsche's distress.

By some estimates, the speculators lost more than €5 billion ($7 billion) on their bets against VW. Investors that suffered losses included Parkcentral Global Hub Holding, which was controlled by the Perot family in Texas and later went out of business for reasons not directly related to Volkswagen, and Greenlight Capital, whose president is David Einhorn, a well-known New York investor.

Another of the short sellers was Adolf Merckle, a German billionaire who made a fortune in generic drugs and had been number 94 in *Forbes* magazine's ranking of the 400 richest people in the world. In January 2009, about two months after the short squeeze, Merckle was found lying in blood-spattered snow next to railroad tracks in Blaubeuren, a village in Baden-Württemburg. Despite his wealth, Merckle and his wife and children had lived in the village in a single-family home with his name on the mailbox. Merckle had stepped in front of an oncoming train.

Porsche bristled at the widespread press reports that it was somehow responsible for Merckle's suicide. Certainly, Merckle had plenty of other reasons to feel despair. The financial crisis had undercut his attempt to take over HeidelbergCement, a German cement producer, by using borrowed money. When HeidelbergCement's shares plunged after the bankruptcy of Lehman Brothers, banks moved to force Merckle to sell Ratiopharm, a pharmaceuticals maker that was the centerpiece of his business holdings. Merckle's bet on a decline in Volkswagen shares appeared to be a last-ditch attempt to raise cash, fend off creditors, and rescue his lifework. When the gamble failed, generating losses estimated to have been as high as €400 million ($560 million), Merckle took his own life.

Even if Porsche was not to blame for Merckle's death, the tragedy cast a pall over the company's financial maneuverings. To be sure, Porsche hardly escaped unscathed. The company reported a loss of €3.6 billion ($5 billion) for the 2008–2009 fiscal year, caused by the Volkswagen options as well as a 24 percent decline in the number of cars sold. The plan to acquire 75 percent of Volkswagen stalled when Porsche's banks, which had enough problems of their own coping with the financial crisis, declined to continue providing financing. Porsche still managed to acquire a slim majority of Volkswagen shares by early January 2009. Given Lower Saxony's blocking minority, the stake was not enough to enable Porsche to dictate terms to Volkswagen. But it gave Porsche more power than any other shareholder. Wiedeking and Härter were forced out in July 2009, but on generous terms, given how much money Porsche had lost during their last year in office. Wiedeking received €55 million ($77 million) in severance plus a pension of more than €100,000 ($140,000) a month. Härter received a severance of about €16 million ($22 million) and a monthly pension of almost €17,000 ($24,000).

With Porsche in a weakened financial state, the Porsche and Piëch families and Volkswagen worked out a deal. Instead of Porsche's acquiring Volkswagen, Volkswagen would acquire Porsche's carmaking operation. Porsche SE, the holding company, would become a vessel for the family shareholding in Volkswagen. Volkswagen would also acquire the family dealer network. The primary sources of Porsche and Piëch wealth would be absorbed into the Volkswagen empire.

At first glance, the agreement looked like a defeat for Porsche. Journalists, including this one, wrote hand-wringing stories about whether the iconic Porsche brand would lose its cachet when it became a mere division of Volkswagen. German newspapers reported that Wolfgang Porsche's voice broke and he was visibly moved when he appeared before weeping Porsche workers on a rainy day in May 2010, as the company was officially becoming part of VW. "The Porsche myth lives and will never die," he said. What many overlooked at the time was that, when all was said and done, the Porsche and Piëch families came out of the agreement with just over 50 percent of Volkswagen voting shares. For all intents and purposes, the Porsche takeover of Volkswagen had succeeded. Without anyone's really noticing, the Porsche and Piëch families had traded a niche, intermittently profitable sports car maker for control of the second-largest carmaker in the world.

And the families had pulled off the takeover largely with Volkswagen's money. Porsche's financial prowess was based on profits made selling Cayenne SUVs manufactured in partnership with VW. Volkswagen was vulnerable to takeover in the first place because, under Ferdinand Piëch's leadership, its share price had languished. Piëch's indifference to stock markets put Volkswagen in play. More than seventy years after Hitler had dedicated the Volkswagenwerk, it belonged to the descendants of the man who had created it. It was one of the most spectacular financial maneuvers ever.

Needless to say, not everyone was happy with the outcome. More than thirty hedge funds that had been burned by the short squeeze sued in U.S. courts. But a federal appeals court ruled that the U.S. courts had no jurisdiction, in part because Porsche and Volkswagen shares were not listed on U.S. stock exchanges. (Some funds also sued in Germany. The cases were still pending as of late 2016.) To be sure, it was hard to shed many tears for the hedge funds. They were operating in a rough neighborhood. As Richard W. Painter, a professor at the University of Minnesota Law School who studied the case, has pointed out, short selling is an inherently risky strategy frowned on, if not banned, in many jurisdictions. It was cheeky of hedge funds to summon protection from U.S. securities laws on shares traded in Germany. That was the equivalent of sitting in New York and betting on a horse race in Stuttgart, then complaining to the New York Gaming Commission that the jockey was corrupt.

But it wasn't just speculators who were unhappy with Porsche's financial sleight of hand. Volkswagen's largest outside shareholder was

Norges Bank Investment Management (NBIM), which administers Norway's oil riches on behalf of the country's citizens. By some measures, NBIM, an arm of the country's central bank, is the largest sovereign wealth fund in the world. In October 2009, the fund addressed a bluntly worded letter to Ferdinand Piëch and the Volkswagen supervisory board expressing dissatisfaction about the terms of Volkswagen's acquisition of Porsche. The deal "was designed to suit the needs of the Porsche controlling families at the expense of Volkswagen and its non-controlling owners," wrote Anne Kvam, the Norway fund's global head of corporate governance, and Ola Peter Krohn Gjessing, a senior analyst.

The Norway fund had a long list of objections. Kvam and Gjessing questioned why, as part of the deal, Volkswagen was buying the Porsche and Piëch dealership network, based in Salzburg. The only justification seemed to be to help the families raise money so that they could use it to purchase more Volkswagen shares and push their stake comfortably above 50 percent. Kvam and Gjessing also criticized the price that Volkswagen was paying for the sports car maker. (The value was hard to pin down because of the complexity of the transaction, but by one estimate VW was paying about €12.4 billion, or $17.4 billion, for the sports car maker.) "The market seems to judge that Volkswagen currently plans to pay a rich price for those assets," they wrote, noting that Volkswagen would also assume Porsche's debts. "This is remarkable at a time when Porsche SE is more in need of a transaction than Volkswagen."

Kvam and Gjessing criticized the supervisory board's decision to finance the acquisition partly by issuing new, preferred shares. When a company sells new shares to raise money, it dilutes the value of existing shares. Shareholders may still come out ahead if the money generated by the share issue is invested in a way that increases the overall value of the company. What annoyed the Norwegians was that Volkswagen was not planning to issue any new voting shares—the kind owned by the Piëch and Porsche families. In other words, the deal was structured to put a disproportionate share of the financial burden and risk on the preferred shareholders, who had no power to block the transaction because they had no voting rights. The whole Porsche transaction suffered from "stealthiness," Kvam and Gjessing complained. They called it "unacceptable."

A public blast of discontent from such a powerful investment fund was extraordinary, but it had no apparent effect on Volkswagen's plans. Because of the complexity of the transaction, Volkswagen did not formally take possession of Porsche's sports car making operations until August 1, 2012. Matthias Müller, a longtime Volkswagen executive with a cool manner, was named the chief executive of Porsche. Fit looking, with neatly trimmed silver hair, Müller was a car guy and a favorite of Piëch's.

The Norwegians were not the only ones who thought the deal was fishy. German law enforcement authorities also began examining the circumstances surrounding Porsche's financial moves. It would take prosecutors in Stuttgart years to build a case, and when they did the timing would be very awkward for Porsche.

CHAPTER 12

Clean Diesel

A MAN STANDS IN HIS KITCHEN, slam-dunking an orange peel into a waste basket. Suddenly he steps back, startled. A blinding white light floods the room. The thump-thump sound of helicopter rotors rattles the window panes. A voice booms from a megaphone. "Put the rind down, sir! That's a compost infraction!" Uniformed men swarm the man's driveway as he tries to flee. The Green Police have collared another eco-offender.

The faux police raid came from an Audi commercial in 2010. The "Green Police" arrest people for installing incandescent light bulbs, overheating their swimming pools, or using plastic bottles. Shot with a fine command of cop drama clichés, the sixty-second spot aired during the Super Bowl, some of the most coveted television time on the planet. It featured hundreds of actors and extras. Audi and its ad agency, Venables Bell & Partners of San Francisco, even got members of the 1970s band Cheap Trick to provide the soundtrack, rerecording their 1979 classic "Dream Police." The new refrain was "Green Police."

Even as the Green Police commercial satirized environmentalist dogma, it offered the implicit promise that people who drove diesels belonged to an elite, morally superior group. The last scene in the spot shows cars backed up at a roadblock where the Green Police are conducting an "eco check." The cops in green shorts notice a driver in an

Audi A3 compact station wagon. "Got a TDI here—clean diesel," says one officer. "You're good to go, sir," says another, waving the A3 out of the line. The Audi swerves around the traffic barriers and speeds away. The other, "dirty" cars must wait.

The effort and expense that went into the commercial showed how determined Volkswagen was to convince Americans that diesel was not only a different kind of fuel but also offered a way to be environmentally virtuous. In Europe, with its astronomical fuel prices, Volkswagen could market diesel on fuel economy. In the United States, where gasoline was cheaper than diesel and much less expensive than in Europe, Volkswagen needed another pitch.

Positioning Volkswagen as a car for environmentally conscious drivers seemed like a clever strategy from many angles. It provided a way to attack archrival Toyota, whose hybrid Prius had become a hit and shown that people would buy a car that lent its owners a green halo. Volkswagen was not in a position to offer competing hybrids, because it had been slow to develop any. But Volkswagen was already a leader in diesel. In the minds of the engineers in Wolfsburg, diesel was the superior technology, anyway. Volkswagen Passats, Jettas, and Beetles were sportier and more fashionable than the homely Prius. If Volkswagen could seem just as green, in their view, it would have a good chance of peeling off some of the worldly drivers most likely to buy a Toyota, or any other import.

Certainly there are people who choose vehicles out of genuine concern for the environment. But cars are also a form of self-expression. The Beetle was a hit in 1960s America not only because it was practical and inexpensive but also because it was a badge of antimaterialism and a form of protest against the ostentatious gas guzzlers churned out by Detroit. Clean diesel was an update of the same pitch.

The green marketing push began soon after Volkswagens and Audis with the new EA 189 diesel motor began appearing on the lots of U.S. dealers in late 2008. Magazine ads used typeface and a minimalist layout reminiscent of the old Doyle Dane Bernbach ads for the Beetle. "This ain't your daddy's diesel," said one. "Diesel has really cleaned up its act," said another. A Volkswagen television spot by the ad agency Deutsch LA showed an elderly lady holding her white scarf in front of the exhaust pipes of a diesel Passat. "See how clean it is?" she tells her friends. Dirty diesel? That's just an old wives' tale, the spot claimed.

The pitch worked well enough to convince people who were environmental activists or professionals, a knowledgeable and discerning group. The Volkswagen owner Cynthia Mackie, of Silver Spring, Maryland, was an ecologist working on international forestry issues. "I bought my Golf 2011 TDI so I could minimize my impact on carbon emissions and overall pollution," Mackie recalled. "So I would not be a hypocrite as a professional ecologist working on natural resource management and on climate change. I did a ton of research and decided the Prius generated more pollution during the manufacturing process. VW convinced me that I made a smart choice."

The Volkswagen buyers were often highly educated. One list of Volkswagen diesel owners included a bankruptcy judge, a married couple who were both professors at Stanford University, an architect who was a graduate of the U.S. Naval Academy, and the executive director of a women's cancer resource center in California. These were not people who would react well to being taken for fools.

Dealers in the United States used the green argument to close sales. They said things like, "You can't go wrong with clean diesel: less emission and more miles per gallon." Or they claimed that TDI exhaust was so clean it "is almost like pool water, drinkable, and safe to inhale." As an added sweetener, many buyers qualified for a $1,300 federal tax credit awarded to cars with lower carbon dioxide emissions.

There was more to the Volkswagen and Audi marketing, later adopted by Porsche for diesel versions of its Cayenne SUV, than just the appeal of being eco-friendly. The engineers in Wolfsburg had tried and, despite the customary threats and verbal abuse from Martin Winterkorn, failed to achieve the same fuel economy as a Prius. But the EA 189 two-liter motor installed in cars exported to the United States came close. And Volkswagen advertised that the official Environmental Protection Agency ratings underestimated actual mileage on the road.

In 2010, for example, a Volkswagen brochure boasted that the latest Jetta with an automatic transmission was "the toast of Europe for its quickness, low emissions, and fuel efficiency—a staggering 38 city/44 highway mpg." That compared to EPA ratings of 30 miles per gallon in urban driving and 42 mpg on the highway, Volkswagen said. Volkswagen drivers could have their cake made from sustainably grown ingredients and eat it, too.

The combination of fuel economy and low emissions appealed to Tony German, a health care administrator in Austin, Texas. He bought a diesel Audi A3, which had the same engine as the Golf, Jetta, and New Beetle. "Both were important to me," German said. He did a lot of driving, and "Austin is a very environmentally conscious community." The fuel economy was real. German said he once drove from Texas to Ohio on two fill-ups. But he didn't know one of the reasons why Volkswagen and Audi diesels got better mileage than the EPA rating. The lean NOx trap that was supposed to neutralize nitrogen oxides needed to be flushed periodically with squirts of extra fuel. By secretly throttling down the emissions controls, the cars squirted less fuel and achieved superior mileage.

Another key element of the campaign to retake America was a new midsize sedan to be made locally. The car would be tailored to American tastes and driving habits and priced to compete against the Toyota Camry and Honda Accord in the market for economical midsize sedans, the sweet spot of the U.S. market. Volkswagen had last made cars in the United States in 1988, when the company closed a factory in New Stanton, in Westmoreland County southeast of Pittsburgh, that had produced Golfs. Although the Pennsylvania factory had been a disaster, Volkswagen needed to try again to manufacture cars in the United States in order to close the efficiency gap with its Asian rivals. Exchange rates were one big consideration. The euro common currency, which was in the process of climbing toward an all-time high of almost 1.60 to the dollar in 2006, raised the price of Volkswagens made in Europe for American customers. But the exchange rate would be largely irrelevant if cars were made and sold inside the United States.

Labor flexibility may also have been an issue. Workers in the United States would not be subject to the laws that gave German workers so much power, and led to the chronic overstaffing that handicapped the efficiency of German plants.

Volkswagen planned to invest one billion dollars in the new plant and hire two thousand people. After Volkswagen began looking for a site in 2007, municipalities around the United States competed intensely for what would be an employment and tax bonanza. Among them was the city of Chattanooga, Tennessee, which offered Volkswagen 1,350 acres of vacant land that had once been home to a dynamite factory. When Volkswagen representatives complained after a visit in 2008 that the long-abandoned site was too overgrown with shrubs and trees for them to judge its suitability, the city brought in a fleet of two hundred earthmovers and wood chippers and cleared away the debris and vegetation in three weeks. It demonstrated how badly Chattanooga wanted the Volkswagen factory, and it worked, to the delight of local politicians. "An auto plant is the holy grail of economic development," said Ron Littlefield, the Chattanooga mayor.

Volkswagen ultimately chose to name the U.S.-made vehicle the Passat, and it had many similarities to the European-made vehicle of the same name. It was powered by a version of the EA 189 motor used in the other Volkswagen diesels. In other respects, the car was a slightly dumbed-down version of its European counterpart. It had a less sophisticated suspension, for example. The German Passats are made for autobahns without speed limits. Americans did not need that kind of over-engineering and it made sense to simplify it in order to keep the price lower.

Still, Winterkorn was keen to offer a vehicle that exuded German engineering expertise. In 2011, the year production began in Chattanooga, he traveled seven times to the United States, according to a sworn statement he made later. Winterkorn took a personal interest in details such as choice of suppliers, and he inspected components himself. His presence sent a powerful signal to the U.S. organization about the importance of their mission. But many American employees, used to a somewhat gentler style of management, were taken aback by his yelling and banging parts around.

At least initially, the U.S. initiatives worked. In 2007, sales of all Volkswagen brand cars in the United States were about 230,000, still far from the 1968 peak of nearly 570,000. In 2011, the U.S. version of the Passat began rolling off the Chattanooga assembly line. By 2012, total sales including gasoline models nearly doubled, to more than 400,000 cars, Volkswagen's best showing since the early 1970s.

Diesels accounted for a substantial proportion of Volkswagen's growth. From 44,000 in 2009, sales of Volkswagen diesels rose to 111,000 in 2013. That compared with fewer than 17,000 diesel passenger cars sold in 2007 by all manufacturers combined. Volkswagen was single-handedly responsible for a sixfold increase in diesel passenger car sales in the United States. While only a little more than a fifth of all Volkswagens sold in the United States were diesels, the proportion was much higher for some larger models. Three-quarters of the station wagon version of the Jetta were diesels, as were more than half of the Tiguan SUVs, probably because the buyers wanted the torque provided by diesel to pull trailers. Diesel cars were also more profitable for Volkswagen. The diesel variants of cars such as the Jetta were priced as much as $6,300 higher than the comparable gasoline versions.

Despite the dramatic increases in U.S. sales that Volkswagen had achieved, Winterkorn's ambitions were far from satisfied. At the christening of the Chattanooga plant in May 2011, he told the assembled dignitaries, including Ray LaHood, the U.S. secretary of transportation, and several U.S. senators, that Volkswagen would sell one million VW and Audi cars in the United States by 2018. The two brands would own 6 percent of the U.S. car market, rivaling Toyota.

The targets were integral to Strategy 2018, the master plan that Winterkorn had announced in 2007 to become the world's largest automaker. Strategy 2018 had evolved in the meantime, undoubtedly in consultation with Piëch, who remained chairman of the Volkswagen supervisory board. The core goal was to sell ten million cars a year worldwide by 2018, becoming what the company's 2009 annual report called "the most successful and fascinating automaker in the world." In addition, Volkswagen vowed to improve its chronically low profit margins, achieving a consistent 8 percent return on sales before taxes by 2018. (In previous years the margin had fallen below 2 percent and risen as high as 7 percent.) And Volkswagen promised "to set new environmental standards in vehicles, powertrains and lightweight construction."

At one level, Winterkorn and by extension Piëch were deploying a familiar management technique. They were establishing demanding, even outrageous targets as a way of focusing the organization and discouraging complacency. There is nothing inherently wrong with that. But top managers must also be careful to set limits on how far employees may go in pursuit of the company's aims. Volkswagen had a code of conduct, which required employees to follow local and international laws, regulations, and treaties. The code acknowledged a duty "to promote mobility in the interest of the common good, while doing justice to individual needs, ecological concerns, and the economic requirements placed on a global enterprise." But the code was little more than lip service if top managers did not give subordinates the resources they needed to achieve the goals legally. And the code was meaningless if top managers did not themselves embrace its principles.

"Stretch goals are very useful," says David Bach, a German-born professor at the Yale University School of Management. "Precisely because they generate a lot of pressure, you have to make sure they are coupled with a clear sense of what the boundaries are. 'We will grow, but we will do so in a way that is true to our brand.' Ensuring those values is more important than reaching the goal. You should never let the goal itself get the better of you. That was just missing at Volkswagen."

This was the context as Volkswagen was rolling out the new Passats made in Chattanooga. Though the diesel versions of the U.S. vehicles had simpler suspensions than their German cousins, they had better emissions systems than the first generation of Volkswagen diesels exported to the United States three years earlier. In fact, the 2012 Passats were equipped with the same kind of technology that Volkswagen had rejected several years earlier when it pulled out of the BlueTec partnership with Daimler. Since then, Volkswagen had acquired its own, similar technology, which it called BlueMotion. Like BlueTec, BlueMotion used so-called selective catalytic reduction, or SCR, a spray of diluted urea to neutralize nitrogen oxide emissions in the exhaust.

The new solution gave Volkswagen an opportunity to phase out the deceptive software needed to pass emissions tests in the United States. Had Volkswagen done so at that point, it is unlikely that anyone would ever have detected the illegal defeat device present in the first generation of vehicles. Volkswagen chose another path.

Diesel was still an unfamiliar technology for most Americans, and Volkswagen executives worried they would be put off by the need to occasionally top off the AdBlue tank. Volkswagen wanted the fluid to last long enough to be refilled by dealers during regularly scheduled oil changes, so there would be no inconvenience to owners. The Passat could not afford to cede any advantage to the popular Toyota Camry and Honda Accord, which ran on gasoline and did not have urea tanks. In addition, U.S. regulations at the time discouraged carmakers from allowing owners to refill the urea tank themselves.

Once again, Volkswagen engineers drew on the precedent set by Audi, which had been using a defeat device to solve a problem with undersized urea tanks since 2008. By the time the people designing the 2012 U.S. Passat confronted the same problem with inadequate urea tanks, the use of illegal work-arounds was already a habit.

DEFEAT DEVICE 2

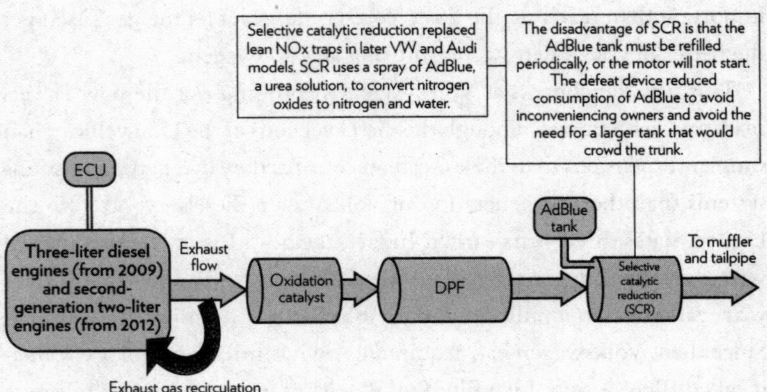

Volkswagen employees who found out about the cheating were told to keep quiet about it. In July 2012, a group of engineers was trying to figure out the cause of mechanical problems with some diesel engines. They discovered software code that made the engine behave differently when it was being tested on rollers, so-called dyno mode. The engineers had stumbled upon the defeat device.

Investigating further, the engineers came to the conclusion that the mechanical problems occurred when the cars ran for too long in dyno mode. That was also when the pollution controls would be operating at maximum effectiveness, putting a strain on the components. The engineers prepared a presentation about what they had found and took it to Heinz-Jakob Neusser, the head of engine development for Volkswagen brand cars since 2011. Neusser's response, according to charges later filed against him, was to tell the engineers to destroy the presentation and to refine the software so that dyno mode would not kick in except when it was needed to fool regulators. The end effect was that the cars polluted more.

Despite the deceit underlying Volkswagen's clean diesel technology, company executives and engineers boasted about how environmentally friendly it was. Volkswagen and Audi engineers spoke at industry events, published academic papers, and gave interviews to enthusiast publications and websites.

Within the engine development community, Volkswagen presented the emissions technology as a breakthrough. In 2013, as Volkswagen was introducing an updated version of the EA 189 diesel known as the EA 288, a group of engineers published an article on the cover of the English edition of *Motortechnische Zeitschrift*, or *Motor Technology Magazine*, a prominent trade publication. The article described how the basic diesel engine could be modified for use in different vehicles while fulfilling the latest emissions standards. The authors were some of Volkswagen's top engine and software developers. In addition to Neusser, they included Jörn Kahrstedt, head of diesel engine development; Richard Dorenkamp, whose title was head of development of lowest emissions diesel engines and exhaust gas after-treatment; and Hanno Jelden, head of drive electronics.

"Thanks to state-of-the-art engineering, the engines of the modular diesel engine system have been developed to meet the future challenges of increasingly strict emissions legislation and increasingly demanding customer requirements," they wrote. "The modular structure of this new design simultaneously provides the opportunity for further reductions in fuel consumption and potential for improved performance."

Elsewhere in the industry, some competitors were skeptical. They couldn't figure out how Volkswagen achieved such clean emissions within the cost constraints of a midrange car. Bob Lutz, who by then had moved from Chrysler to become a vice chairman of General Motors, wanted to offer competing products and badgered his engineers to match Volkswagen's apparent diesel efficiency. But they never could, even though GM bought components from many of the same suppliers as Volkswagen, like Bosch and Continental. "Volkswagen is producing in the United States," Lutz said to GM engineers at one meeting, "and you're telling me GM, with all our resources, we can't do it?"

"I was beating up on 'em pretty hard," Lutz said. "Nobody could figure out how they were complying, and their engine technology is very similar to everybody else's." But Lutz never shared his doubts with any regulators. "The unwritten rule is you don't throw another company under the bus."

The deployment of defeat devices at the same time that Volkswagen trumpeted its commitment to the environment seems like an extreme case of corporate cognitive dissonance. Yet, according to several former executives, Volkswagen's top managers sincerely thought of themselves as leaders in sustainable transportation. It was true that Volkswagen diesels produced less carbon dioxide than comparable gasoline motors, and contributed less to global warming. In 2012, Greenpeace activists unfurled a banner from the ceiling of an auditorium in Hamburg where Winterkorn was giving a speech to shareholders. "Honest Climate Protection Now!" read the banner, which blocked a giant video projection of Winterkorn. At a supervisory board meeting, Winterkorn and Piëch seemed genuinely wounded by the protest, which accused the company of making only token efforts to minimize the effect of its vehicles on the environment. "They said, 'But we're clean!'" recalled Jörg Bode, a Lower Saxony politician who was a member of the supervisory board. "If they were just acting," he said, "they did a pretty good job."

In Europe, use of software to mask emissions output was a misdemeanor, with no grave penalties. The regulations in the European Union and United States prohibiting defeat devices used almost identical wording. But EU law made no provision for penalties. The worst that could happen would be that Volkswagen would be forced to recall cars to make them compliant. Oversight was also weak. Emissions testing was done by outside contractors hired by car manufacturers. The contractors were ostensibly independent, but had little incentive to take a hard line against the companies paying them. Enforcement of emissions rules was left to national authorities. In Germany, the oversight agency was the Kraftfahrt Bundesamt, which was under the supervision of the Ministry of Transportation. The ministry was well-known for being friendly to the auto industry.

European regulations provided carmakers with a loophole that made it easy for them to pass emissions tests without resorting to illegal cheating. The manufacturers were allowed to dial back pollution controls if there was a risk of damage to the motor. Carmakers were not required to disclose when they took advantage of this loophole; in the United States, by contrast, manufacturers needed permission from the Environmental Protection Agency to deactivate emissions equipment under certain circumstances. As Volkswagen well knew, almost all other European carmakers made broad use of this loophole. For example, virtually all of them programmed their diesels to reduce the amount of exhaust gas recycled back into the motors when the outside air temperature was lower. The strategy reduced the risk that the systems would clog, but also led to higher nitrogen oxide emissions. Automakers could argue that they were protecting the motors. But they also knew that regulators in Europe tested cars at twenty degrees Celsius, or sixty-eight degrees Fahrenheit, and would never notice what engines did under cooler conditions. Europe made it so easy to undermine emissions standards legally, in fact, that defeat devices were completely superfluous.

But there was no such loophole in the United States. Cars were also tested at lower temperatures, and penalties could be severe.

The leeway for anyone inside Volkswagen to resist the demands of top management had, if anything, grown narrower. In 2013, Volkswagen sold 9.7 million vehicles, and was close to meeting its sales target ahead of schedule. The closer Volkswagen came to surpassing Toyota as the largest carmaker in the world, the more powerful Winterkorn and Piëch became. Winterkorn was by 2013 one of the best-paid managers in Europe, with annual compensation of €15 million, or $19 million. He had a fleet of company jets at his disposal, including an Airbus A319 designed to hold up to 160 passengers. During flights executives jostled to sit as close as possible to him in the front of the plane, according to a former executive who witnessed the behavior. The closer a manager sat next to the boss, the higher his status.

Winterkorn was treated like a regent when he traveled the far-flung Volkswagen empire. Managers at a plant in Brazil once repainted the wall of a factory just because Winterkorn was scheduled to walk by it. He was a presence in a way that other auto executives were not. Norbert Reithofer, the chief executive of BMW, traveled alone or with an aide or two. That was not Winterkorn's style. His imperial bearing was on view in 2011 when an anonymous person with a video camera captured him touring competitors' displays at the Frankfurt motor show. Trailed by an entourage, Winterkorn stopped at a Hyundai i30 compact, which was making inroads into the European market. Winterkorn removed a pen-sized device from the pocket of his double-breasted suit and used it to examine the sheet metal gaps in the Hyundai's hatchback. Ignoring a blond car-show model, he paced around the car with a stern expression, then lowered his stocky frame into the driver's seat. Other Volkswagen executives clambered into the other seats and another crouched next to the open driver's side door attentively. Winterkorn ran a finger along the interior plastic, then fiddled with the adjustable steering wheel. "It doesn't clank," Winterkorn said gruffly to a member of his auto design team, speaking in his native Swabian dialect, which sounds rustic to Germans from other regions. "BMW can't do it. We can't do it. They can."

"We had a solution, but it was too expensive," the designer answered nervously.

Winterkorn was not appeased. "How come they can?" he said.

Rank offered no immunity from Winterkorn's ire. Every Tuesday, top Volkswagen managers and division heads gathered at the top of the brick high-rise, with its hood ornament–like VW insignia on top, which overlooked the factory complex in Wolfsburg. The managers might be heads of Volkswagen brands like Audi or Skoda that were carmaking empires unto themselves, with tens of thousands of workers. But that status offered no protection on Tuesdays. Anyone who had failed to meet his goals (they were all men) could expect merciless criticism from Winterkorn in front of his peers, according to several people who witnessed the meetings. Managers who were favorites one week could fall suddenly from grace the next. Sometimes they learned they had been demoted or dismissed not from Winterkorn or a colleague but by reading about it in a German business publication, like *Manager Magazin*, which had somehow been tipped off.

Piëch, then in his midseventies, remained a dominant presence in the company and another reason for engineers to fear the consequences of failure. Piëch demonstrated how thoroughly he controlled the supervisory board by arranging for his wife, Ursula, to be named a member in 2012. She was the third woman on the twenty-person board. Ursula Piëch had little if any formal business training. But because she was only in her fifties, and would inherit her husband's wealth, she would be in a position to protect his legacy even after he was gone. Snarky commentary in the German media about the obvious nepotism was ignored. The business daily *Handelsblatt* pinpointed what was probably Ursula's greatest qualification for the job. "This clever woman has achieved something that no one else has managed: the grumpy, deeply skeptical Ferdinand Piëch has granted her his trust." Ferdinand Piëch continued to pay close attention to Volkswagen product development. He attended auto shows and took part in company events in South Africa and other locations where top executives tested and critiqued the latest prototypes.

Thanks to Volkswagen's success in marketing "clean diesel," by 2013 the EA 189 motor had become ubiquitous. There were 1.2-liter, 1.6-liter, and 2-liter versions. (Only the 2-liter was exported to the United States.) Millions had been produced in factories in Salzgitter, Germany, not far from Wolfsburg, as well as in Polkowice, Poland, and Györ, Hungary. The motors were found not only in Volkswagen and Audi brand vehicles but also in two of the company's other passenger car brands: SEAT, based in Spain, and Skoda, based in the Czech Republic. SEAT and Skoda, which had body styles different from Volkswagen's but shared components with them, were not sold in the United States. But they were available almost everywhere else in the world. The EA 189 was an exceptionally versatile motor, used in compacts like the Golf and Jetta as well as in light trucks and smaller SUVs like the Tiguan. The vast majority of cars with the motor were in Europe, but they were also sold in Latin America and Asian countries such as South Korea.

The EA 189 diesel was a prime example of Piëch's strategy of sharing components among as many vehicles as possible. The motor was also an example of the risks of that strategy. All the motors shared an engine control unit illegally programmed to foil emissions tests. By 2013, according to evidence presented by investigators in the United States, Volkswagen had deployed defeat devices not only in the EA 189 but also in three-liter motors used by Audi and Porsche. In addition, the company had programmed the onboard diagnostics system to provide false data to prevent the defeat devices from being detected by inspectors or mechanics. Each car on the road was a violation waiting to be discovered, and the more engines Volkswagen produced, the greater the chances of exposure.

CHAPTER 13

Enforcers II

AMONG A SMALL CIRCLE OF EXPERTS and activists in Europe and the United States, there was considerable skepticism about the "clean diesel" claim. But, with diesel accounting for half the passenger cars sold in Europe and gaining ground in the United States, no one paid much attention to the people warning of the downside. Auto industry lobbyists had done a good job persuading most political leaders in Europe that diesel was environmentally friendly, which was half true. Diesel vehicles produced less climate-changing carbon dioxide. But they spewed far more poisonous nitrogen oxides than gasoline cars did, as well as very fine, carcinogenic soot particles.

There is overwhelming scientific evidence that nitrogen oxides are extremely harmful to human health and sometimes deadly. Nitrogen oxides, or NOx, are made up of various combinations of nitrogen and oxygen atoms. About 80 percent of the earth's atmosphere is diatomic nitrogen, which consists of two nitrogen atoms bound together. Diatomic nitrogen is harmless and indeed essential to life, but it turns evil inside a diesel engine. The high combustion temperatures inside the cylinders of a diesel motor separate the atoms into single nitrogen atoms, some of which combine with oxygen to form an array of harmful molecules, in particular nitrogen dioxide. All motors that run on fossil fuels produce nitrogen oxides, but diesel engines are especially prolific producers of nitrogen oxides because their fuel ignites at much higher temperatures than that in gasoline engines.

Most of what comes out of a tailpipe takes the form of single nitrogen atoms. These react quickly with ozone and other substances in the atmosphere to form nitrogen dioxide, the gas most damaging to human health. When inhaled, nitrogen dioxide reacts with the fluid lining the lungs, causing inflammation and allergic reactions that can lead to asthma or, for someone already suffering from asthma, to have an attack. Children are especially at risk. Studies have found a correlation between areas with high nitrogen dioxide concentrations and the percentage of children who develop asthma.

There is also evidence that exposure to nitrogen dioxide pollution can cause problems with the cardiovascular system, contribute to diabetes, lead to chronic bronchitis, and cause cancer. Studies, cited in the EPA's exhaustive report *Integrated Science Assessment for Oxides of Nitrogen—Health Criteria*, have found a correlation between spikes in nitrogen dioxide levels in city air and emergency room visits by people suffering a heart attack. Overall death rates and infant mortality are greater in areas with high nitrogen dioxide concentrations, though it is difficult to determine what other pollutants might have also played a role. Some research has suggested that nitrogen dioxide can impede the growth of fetuses. Likewise, studies have suggested a link between nitrogen dioxide and lung cancer.

Besides being poisonous in its own right, nitrogen dioxide is the main cause of the smog that plagues large cities. Exposed to sunlight, nitrogen dioxide reacts with other substances in the atmosphere to form ground-level ozone, or smog, which also causes asthma and other respiratory problems, contributes to global warming, and harms some kinds of plants. Emissions from diesel motors produce other combinations of nitrogen and oxygen that have ill effects on people and the environment. Among them is nitric oxide, a cause of acid rain. Another is nitrous oxide, also known as laughing gas, which may be useful in dentists' offices but is a potent greenhouse gas when released in the atmosphere. Nitrous oxide rises into the upper atmosphere, where it can take a century to break down. Until then, it helps trap heat and warm the planet. Pound for pound, nitrous oxide is 300 times as potent as carbon dioxide in its effect on the climate, according to the EPA.

Motor vehicles are the largest single source of nitrogen oxides in populated areas. In urban areas in the United States, tailpipe emissions account for anywhere from 40 percent to 67 percent of nitrogen oxide pollution, the EPA has found. Not surprisingly, the highest concentrations of nitrogen oxides are near highways. About a fifth of the U.S. population lives within one hundred yards of a major road. In Los Angeles, 40 percent of residents do. Children tend to suffer the highest exposure because they spend more time outdoors than adults. There is also evidence that poor children are most likely to suffer exposure to nitrogen oxides because schools in low-income neighborhoods are often in heavily trafficked areas.

Europe has a much more serious problem with nitrogen oxides, one that makes the pollution in the United States seem almost trivial by comparison. Diesels account for about half of all vehicles sold in Europe, which is more densely populated and urbanized than the United States. Air pollution of all kinds was responsible for a staggering 403,000 premature deaths in 2010, according to European Union figures. About 30 million Europeans are exposed to quantities of nitrogen dioxide that exceed official standards.

Despite those alarming statistics, diesel was popular with European drivers because it was cheaper. The car industry convinced legislators in many countries to set a lower fuel tax on diesel, so it was typically about twenty euro cents per liter less than gasoline. Hybrids, the main competing technology for drivers concerned about fuel economy and carbon dioxide emissions, were slow to catch on in Europe. Toyota, the pioneer in hybrid technology, had never been as successful in Europe as it was in the United States. Its market share was in the low single digits, in niche player territory. European manufacturers, including Volkswagen, offered hybrid models, but without enthusiasm and with little promotion. Amid the dominance of diesel, the people sounding warnings about its health effects were regarded as gadflies and spoilsports.

Among the naysayers was Axel Friedrich, a retired civil servant who had worked at the Umweltbundesamt, the German equivalent of the Environmental Protection Agency. Unlike the EPA, the Umweltbundesamt was not directly responsible for enforcing vehicle emissions rules. That was the province of the Kraftfahrt-Bundesamt, the motor vehicles agency that, among other things, keeps track of how many points German drivers accumulate for traffic infractions. The Kraftfahrt-Bundesamt was part of the German Transport Ministry, which had a reputation for being protective of the German auto industry. But the Umweltbundesamt, part of the Interior Ministry and less beholden to carmakers, was responsible for overall air quality, and had a brief to weigh in on issues of vehicle emissions.

Friedrich was the former head of the Umweltbundesamt's Traffic and Noise Department, where he developed a reputation for being a thorn in the auto industry's side. Air quality was Friedrich's passion. Early in his career, he coauthored a book, *Everything You Ever Wanted to Know about Air Pollution Control*. Friedrich was among those who noticed that pollution levels in cities were not dropping as much as they should have been, considering how much stricter emissions rules had become. He was at the forefront of those asking why. Part of the answer, his agency discovered, was simple: the vehicle manufacturers were cheating. In 2003, Friedrich's department published a study showing that European truck makers were using the computers inside diesel engines to evade emissions regulations. In tests conducted with counterparts in Austria, Switzerland, and the Netherlands, researchers had shown that trucks emitted more nitrogen oxides in laboratory tests if technicians deviated from the simulated driving pattern used by regulators. The study did not use the term "defeat device," but the clear implication was that the trucks were programmed to behave only when the engine software detected that the standardized regulatory test was underway. The rest of the time, anything went. The findings helped explain why, despite progressively stricter truck emissions standards, nitrogen oxide levels in cities had actually gotten worse.

In effect, Friedrich and his team had discovered wrongdoing on the same scale as the emissions cheating by truck engine makers in the United States—the scandal in 1998 that had led to a $1 billion settlement. But there were virtually no consequences for the truck industry in Europe, even though its behavior was illegal. European laws made no provision for the kind of penalties that the EPA was able to impose on the American truck engine makers. The report by the Umweltbundesamt did, at least, prompt European regulators to eventually require that trucks be tested on the road as well as in laboratories, to make cheating more difficult.

Friedrich's work did not render him popular with vehicle manufacturers, nor did his stubbornness. The German magazine *Stern* described him as an "irksome but well-informed critic of the auto industry." Gray-haired and wiry, Friedrich considered himself a pragmatist who knew what was politically possible and what was not. He drove an ancient—but fuel-efficient—Fiat 500, which could be seen as an expression of his contrariness. He admitted, "I'm a bad loser. When I start something I finish it."

Indeed, stubbornness is required to take on the auto industry in Germany. Cars are the country's biggest export and employ about 800,000 people. The manufacturers are adept at leveraging this economic clout, maintaining close connections to political insiders and hiring ex-politicians as lobbyists. For example, Matthias Wissmann, the president of the German Association of the Automotive Industry, was a former federal transport minister. A lawyer and smooth political operative, Wissmann served in the cabinet of Chancellor Helmut Kohl at the same time as Angela Merkel, who later became chancellor herself. Wissmann addressed Merkel by her first name, which Germans normally do only with close friends.

Volkswagen's political clout was particularly potent. The state of Lower Saxony, where Wolfsburg is located, is a launching pad for national politicians. They included a chancellor, Gerhard Schröder, and Christian Wulff, the federal president from 2010 to 2012. Both were former prime ministers of Lower Saxony. Both had been members of the Volkswagen supervisory board. The carmakers also had a history of taking care of unemployed politicians who had been accommodating. After Sigmar Gabriel, another prime minister of Lower Saxony, was voted out of office in 2003, Volkswagen awarded a contract to a consulting firm in which he owned a stake. Gabriel later succeeded in reviving his political career and became chairman of the Social Democrats and, in 2013, economics minister in a coalition government with Merkel's Christian Democrats. Volkswagen had lots of friends in Berlin.

Though Germany had a strong environmental movement, and Chancellor Merkel boldly pushed for a ban on nuclear power following the Fukushima disaster in 2011, the economic prerogatives of the auto industry often took precedence. In 2013, when the European Parliament was getting close to approval of strict new limits on carbon dioxide, Wissmann wrote a letter to Merkel on behalf of the auto industry complaining about the "poorly considered" standards. Wissmann warned that the new CO2 limits would be particularly hard on makers of high-performance luxury cars—a reference to Audi, Mercedes, and BMW—which accounted for 60 percent of auto industry jobs in Germany. The letter on industry association stationery was addressed formally to the "sehr geehrte Frau Bundeskanzlerin" (very honored Frau Chancellor), but Wissmann wrote by hand in the margin, "Dear Angela." Following German objections, the CO2 proposal was watered down in Brussels.

Friedrich remained a leading diesel skeptic even after his retirement from the Umweltbundesamt in 2008. For example, he became one of the founders of the International Council on Clean Transportation, which was originally conceived as a network linking environmental officials around the world so that they could exchange information and coordinate their activities. The ICCT also included a number of former EPA officials, all of whom shared a certain frustration with what they regarded as the auto industry's influence over regulation. If the automakers coordinated their political activities internationally, Friedrich and other ICCT founders reasoned, so should the regulators.

Particularly frustrating to Friedrich was the carmakers' resistance to stricter nitrogen oxide limits in Europe. He regarded nitrogen oxide pollution as a form of physical assault. But the carmakers insisted that more stringent limits, like those the United States had already imposed, would be too expensive. Yet BMW, Mercedes, and especially Volkswagen were selling diesels in the United States and managing to meet the stricter American standards. They were achieving in the United States what they said they couldn't achieve in Europe. Friedrich's assumption was that the German automakers were deploying technology in America that, to save money, they did not want to install in European vehicles. "We wanted to show it was working in the United States," he said. "That was the idea."

Not only were American pollution rules more strict, allowing less than half as much nitrogen oxide emissions as those in Europe, but so was enforcement. In Europe, it was practically unheard of for carmakers to be penalized for clean air violations, and European regulations lacked specifics. In the United States, the EPA provided automakers with extensive documentation on what was allowed and what wasn't, and there was ample precedent. In Europe, vague regulations created a huge gray area that carmakers exploited. Diesels on U.S. roads were more likely to be compliant with the spirit of clean air rules—or so the ICCT researchers thought. "We really tried to do good analytical work," said John German, a senior fellow at the ICCT who had worked for Chrysler and Honda as well as the EPA. "We realized there was a hole in our data—we had no data from the US." The expectation, German said, was that the ICCT would go back to Europe and be able to say of the carmakers, "They can do it there. Why can't they do it here?" In 2012, the ICCT went looking for a facility in the United States capable of proving that assertion.

CHAPTER 14

On the Road

DAN CARDER AND HIS TEAM at West Virginia University had worked with ICCT before. Francisco Posada, a senior researcher at the ICCT who was overseeing the project, was a WVU graduate. Still, Carder had somehow overlooked the ICCT call for proposals; he heard about it only from someone at Horiba, the Japanese maker of portable emissions-testing equipment. In 2011, a year earlier, Carder had become director of the WVU emissions lab and, as usual, was on the lookout for grants and contracts to keep the shoestring CAFEE operation going. Carder asked Hemanth Kappanna, a native of Bangalore, India, who was studying for his doctorate, to draft a proposal for the ICCT job.

The application that Kappanna prepared, with important input from Greg Thompson, the associate professor who had been a key figure in CAFEE since its founding, called for the university team to test a variety of European diesel models and emissions technologies at a cost of $200,000. The ICCT awarded WVU the job, but asked Carder to scale back the scope of the testing so that it could be done for $70,000. It wasn't a lot of money, on the low end of the contracts that CAFEE usually got. A BMW, Mercedes, or Audi could easily cost that much with options. But it was something. Carder took it.

CAFEE had wide experience testing trucks, but except for a few small-scale student projects had not done much road testing of passenger cars. There was no demand, because of the prevailing view that car emissions were not a big problem. As a result, the ICCT contract presented some technical and practical issues. The first was finding the cars to test. The contract allowed Carder and his crew to choose from among any of the European diesel cars sold in the United States, which included Mercedes, BMW, Volkswagen, and Audi. ICCT wanted to test the two prevailing technologies for treating nitrogen oxide emissions, the lean NOx trap and the urea-based SCR system. Volkswagen and Audi used both technologies. The first generation of cars sold in the United States that had the EA 189 motors, beginning with the 2009 models, had lean NOx traps, which were cheaper and did not require periodic refills of the urea chemical solution. Cars with that technology included the VW Jetta, VW Golf, and Audi A3. In 2012, Volkswagen had introduced SCR systems in versions of the Passat designed for the U.S. market and built at the company's brand-new factory in Chattanooga. SCR systems were considered more effective, but, as previously noted, had one disadvantage. The tank containing the urea chemical solution, sold commercially as AdBlue, needed to be refilled periodically. If the driver ignored warnings to refill the fluid, eventually the car would not start.

Carder and Thompson were ignorant of the political agenda behind the ICCT contract. They were just supposed to do the technical work. "We never set out to get crosswise with anyone," Carder said later. "We were just kind of doing our jobs."

Carder originally planned to do the road testing around West Virginia, close to home and at lower cost. But it proved too difficult to get hold of the cars. Local rental agencies didn't have the European diesels, and the WVU team couldn't find any owners in the area willing to lend their cars. Arvind Thiruvengadam, a graduate student originally from Chennai, India, who was working on Carder's team, called Volkswagen in Auburn Hills, Michigan, to ask whether the company would provide vehicles. VW refused.

Carder decided instead on California, where CAFEE had often done work in the past and where there would be a bigger pool of vehicles. The decision proved to be crucial in the sequence of events that led to Volkswagen's exposure. The California Air Resources Board has a reputation for toughness that corresponds to the state's air pollution problem. The state has 25 million vehicles for a population of 34 million. Six of the ten U.S. cities with the worst air pollution are in California; Los Angeles has the worst air of all, according to the American Lung Association. Because of the state's continuing struggle to clean up the problem, California has stricter air pollution guidelines than the federal government. CARB benefits from a bipartisan political consensus in favor of rigorous clean air enforcement, in contrast to the EPA, often a political piñata in Washington. Within the auto industry, CARB is respected and sometimes feared for its expertise.

West Virginia University and CARB already had close relations. Alberto Ayala, the head of CARB's emissions testing laboratories, had been on the WVU faculty in the late 1990s and was still an adjunct professor. Ayala, who has a doctorate from the University of California at Davis, comes across as a scientist who chooses his words carefully and does not draw conclusions without data. He has a thin mustache and keeps his hair shaved close. Normally guarded, he displays flashes of anger. Ayala's original specialty was aeronautics, but he switched to air pollution after cuts in the U.S. defense budget made aerospace jobs in California scarce.

Ayala had already been paying close attention to diesel passenger cars. He was intrigued with the German automakers' diesel technology because of its potential to help lower carbon dioxide emissions. At the same time, Ayala was in close contact with experts from the European Commission and the ICCT. He was aware that nitrogen oxide levels in European cities were higher than they should have been. He and others at CARB had some doubts about whether diesel was as clean as advertised.

Nitrogen oxide pollution is an especially big concern for California. When NOx comes out of the tailpipe of a truck or car, it quickly reacts with sunlight and gases in the atmosphere to create smog. Los Angeles, with its cloudless skies and basin-like topography, is an ideal place for smog to brew. In fact, the CARB lab in El Monte is named for Arie J. Haagen-Smit, the Dutch scientist and California Institute of Technology professor who proved that smog was caused by the interaction of automobile exhaust and sunlight. His work created the scientific basis for air quality regulations. Haagen-Smit, who died in 1977, became CARB's first chairman when the agency was created in 1968, two years before the EPA.

Ayala saw that West Virginia University's research could advance CARB's knowledge base. He offered to let Carder's team borrow the labs in El Monte, east of downtown Los Angeles, where the agency tests passenger cars. The labs are located in a drab, one-story concrete and cinder block building from the 1970s ringed by an iron fence with barbed wire on top. Inside are four test bays, garage-like spaces with concrete floors and large metal rollers in the floor that allow the cars to spin their wheels at high speed without crashing into the wall. Metal pipes connect to the vehicles' tailpipes in the collection of exhaust gases. The bays have large fans that blow air at the cars' front ends, in order to simulate highway conditions as closely as possible. Ayala allowed the WVU graduate students doing the fieldwork—Kappanna, Thiruvengadam, and a Swiss doctoral candidate named Marc Besch—to use the facilities to run standard emissions tests that would be needed to provide a basis for comparison with road tests.

California also offered Carder's crew members a larger pool of vehicles. They soon found the diesels they needed. They were able to rent a 2013 BMW SUV and a 2012 VW Jetta from local rental agencies, and found an owner willing to lend his 2012 Passat. They had hoped to test a Mercedes diesel as the third car, but at the last minute the owner demanded more money. They opted for the second Volkswagen instead. The Jetta had about 4,700 miles on the odometer when the tests began; the Passat and BMW, about 15,000 miles each. They were broken in, but not old enough for the emissions equipment to be worn out. All three were station wagons, which had room for the equipment in back.

When the WVU team members picked up the Passat, they had a conversation with the owner that didn't seem important at the time, but in retrospect should have been a red flag. The Passat owner had never refilled the urea fluid needed to make the car's SCR system function. In fact, he had no idea there was a separate AdBlue tank that had to be refilled. The owner's ignorance provided a clue that AdBlue consumption was curiously low. "That should have been an 'aha' moment," Carder recalled.

The next problem for the WVU team was to figure out a way to power the emissions analysis gear while the test was underway. The plan agreed upon with the ICCT called for the testing of cars over longer stretches. But the Horiba equipment the team intended to use was designed for relatively short sessions of less than an hour and had limited battery capacity. (The equipment couldn't draw power from the car being tested, because the extra load on the engine would throw off the results.) The solution was to buy portable gasoline generators at hardware supply outlets and bolt them to the back of the test cars. The generators were noisy and smelly and frequently ran out of fuel. But they provided the necessary power.

All three cars passed the standardized tests in CARB's El Monte lab. But once the three students got out on the highway, they quickly realized that something about the Volkswagens was odd. The Jetta, the one with the lean NOx trap, belched unusually high levels of nitrogen oxides. The Passat with the SCR system was a little better, but still way over the legal limits. And it wasn't just that the emissions were excessive. The behavior of the whole system was funny. For example, normally there would be a big drop in emissions after the car warmed up. But there wasn't. The BMW, on the other hand, did not show such an extreme deviation between emissions on the road and those in the lab. For the most part, the BMW emissions controls worked.

The students drove a circuit in and around Los Angeles that, as it happened, was the model for one of the simulated driving cycles used by the EPA. It included city and highway driving as well as a trip to Mount San Antonio, a peak outside Los Angeles popularly known as

Mount Baldy. Though students, the testers were experienced. Kappanna, for example, had earned an undergraduate degree in mechanical engineering in India and went to WVU on the recommendation of a friend who studied there. (Kappanna jokes that his only prior knowledge of West Virginia came from hearing the John Denver song "Take Me Home, Country Roads," in Bangalore bars.) Kappanna had been involved in diesel truck testing while getting his master's degree at WVU. After graduation, he went to work at Cummins, but he lost his job in 2008 during a round of corporate cost cutting and decided to go back to WVU to get his doctorate. Besch had studied automotive engineering at the technical university in Biel, Switzerland, where his professor was one of Europe's leading emissions experts. Thiruvengadam had earned an undergraduate degree in mechanical engineering at the University of Madras in India before pursuing his doctorate in West Virginia. He had already contributed to numerous published research papers on subjects like diesel particle emissions or the greenhouse gas output of heavy trucks. He later became a tenure-track associate professor at West Virginia University.

The three doctoral candidates drove the circuits around LA numerous times and also did tests in San Diego and San Francisco. They drove at different speeds, on divided highways, in rush hour traffic, in the foothills around LA. Thiruvengadam and Besch took the Passat from Los Angeles to Seattle and back, a 4,000-mile trip. The portable generators, not meant to be bounced around in the back of a car, broke down often and had to be replaced. "It was stinking, it was painful in the ears, it was stressful," Kappanna recalled. The measuring equipment was finicky, too. Wires came loose, sensors failed. Once Besch and Thiruvengadam spent a late night repairing the gear in the parking lot of a Lowe's home improvement store in Portland, Oregon. In northern California, they were questioned by a police officer suspicious of the bizarre gear protruding from the back of the test car.

The students kept expecting the Volkswagen emissions to average out to be close to the official standards once all the data was collected. It is normal for a car's pollution output to fluctuate according to weather conditions or when the vehicle is climbing a hill. But the Volkswagen emissions remained stubbornly high.

None of the students suspected deliberate wrongdoing by Volkswagen. Their first thought was that something was wrong with their own equipment. But everything checked out. Another possible explanation was that Volkswagen had worked out a deal with the regulators allowing it to exceed the standards under certain conditions. Carmakers often negotiate exceptions to the rules, and sometimes pay the EPA money, which is used to offset the effects of the excess emissions in some other way—for example, by retrofitting older school buses' engines to pollute less. Such agreements are considered trade secrets and are not made public, so there was perhaps an explanation that the students wouldn't have known about. When Carder saw the data, he attributed the high emissions to a design or engineering flaw in the cars, not deliberate wrongdoing. He expected, at worst, that Volkswagen would have to recall some vehicles. "I was thinking more, 'This is going to cost Volkswagen some money,'" Carder said. "They're going to have to go back and fix some things." He couldn't imagine that, after all the grief the truck makers suffered after being caught using defeat devices in 1998, "someone would purposely try to do something like that."

By mid-2013, the data collection was complete. Carder and his team took a while to compile the information; it wasn't at the top of their list of priorities. Greg Thompson, the senior academic in the group, was pleased that the research had produced data that helped Besch complete his doctoral thesis on diesel particulate emissions. That was "the biggest value we got out of this," Thompson said. Or so it seemed at the time.

The 117-page research paper they produced was entitled "In-Use Emissions Testing of Light-Duty Diesel Vehicles in the United States." Thompson was listed as the principal investigator, with Carder, Besch, Thiruvengadam, and Kappanna as coinvestigators. Their report didn't specify what cars had been tested, referring to them simply as vehicles A, B, and C. Amid charts and paragraphs of technical data was the revelation that Vehicle A—the Jetta—emitted up to thirty-five times the legal limit of nitrogen oxide on the road. The car never fell below regulatory limits in driving conditions, and for fully half the time the car was on the road it was at least twenty times over the limits. Vehicle B—the Passat—was up to twenty times over, and on the road only rarely complied with pollution laws. Vehicle C, the BMW, was "vastly different," the study said. It exceeded pollution limits only in uphill driving, and then only by a factor of ten. The discrepancy between the BMW and the Passat was especially puzzling. Both used essentially the same SCR emissions technology to control nitrogen oxides. One explanation, the WVU authors speculated, was that the Passat might be deploying a "different diesel exhaust fluid injection strategy," perhaps to reduce consumption of the urea solution and therefore the need for refills. The paper also noted that, "interestingly," the Jetta and Passat had performed perfectly when tested on rollers in the CARB lab in El Monte.

Besch presented a summary of the paper at an annual conference for emissions experts in San Diego at the end of March 2014. It was not exactly a convention of tree huggers. The gathering, known as the Real World Emissions Workshop, was sponsored by the Coordinating Research Council, which is funded by the petroleum industry and the major automakers, including Volkswagen. Besch remembers about two hundred people being in the audience at the conference venue, a Hyatt Regency in San Diego with views of the Pacific. Among them were officials of the EPA and CARB and at least two people from Volkswagen's emissions compliance department. Besch still did not reveal which cars had been tested, but it was not hard for the experts in the audience to figure out. The paper provided detailed technical specifications as well as photographs of the vehicles' rear ends fitted out with testing equipment. During the break that followed Besch's talk, several people approached Carder to tell him they knew who the manufacturer of Vehicles A and B must be: Volkswagen. Some commented that the company would have to do a recall, according to Carder, but no one spoke of a defeat device, much less a scandal. "I don't remember people kind of taking it in and saying, 'Wow, this is really big,'" Carder recalled.

At least one person already suspected that the strange test results produced by the Volkswagen diesels were the result of deliberate deception.

German, the senior fellow at the ICCT who had been involved in hiring WVU, had stayed in touch with Carder and his team during the testing and received continual updates on their findings. With his experience in the auto industry and the EPA, German recognized the telltale signs of a defeat device. If the excess emissions had been the result of a malfunction, the onboard diagnostics software would have reported a problem. A dashboard warning light would have gone on. But it didn't. German said nothing about his suspicions, not even to his colleagues at the ICCT. In German's mind, accusing a carmaker of cheating on emissions by using a defeat device was like accusing a priest of child abuse. "It's not something you do unless you're sure."

The job done, Carder, Thompson, and the three students went back to West Virginia and moved on to other projects. But the paper had generated a bigger reaction than they realized.

Volkswagen officials in the United States and Germany knew about WVU's work even before the paper was published. Ayala had paid close attention to the data being collected by Dan Carder and his team, who were often at the CARB labs. Ayala, in turn, had been keeping Volkswagen informed, in line with his philosophy of maintaining open lines of communication with the car manufacturers that CARB regulates. "They knew that we were doing this," Ayala said. Ayala met with Volkswagen representatives both in El Monte and in Sacramento, the state capital that is his main base of operations. "We were finding things that just didn't make sense," Ayala recalled. Though bothered by the results, he didn't suspect wrongdoing. He still thought the excess emissions were due to a particularly thorny technical problem. Relations with Volkswagen were cordial. Volkswagen began doing its own testing in response to the WVU results. Ayala and company experts compared technical notes about the methodologies they were using. That kind of exchange of information is not unusual. "My premise was, 'Let's work together to solve the problem,'" Ayala said. He still trusted Volkswagen.

Ayala even mentioned the WVU project during a routine visit to Wolfsburg for several days in December 2013. CARB officials periodically visit carmakers to discuss how well air quality policy is working.

Ayala, who by then had been promoted to deputy executive officer of CARB, brought up the WVU study. "We were talking openly about the things that we were finding, the fact that we need to do a little more research," said Ayala, who declined to name the Volkswagen executives he met in Wolfsburg. According to Ayala, the Volkswagen people replied, "Sure, we are also looking at that."

After WVU presented its report in March, there was a flurry of anxious e-mails among executives at the highest levels of the company. Several requested copies of the study, including Michael Horn, who was president of Volkswagen Group of America, and Christian Klingler, a member of the Volkswagen management board in Wolfsburg who was responsible for sales. Considering that Horn and Klingler were both marketing guys, not engineers, their interest in a highly technical academic paper from a not especially prestigious state university was noteworthy.

Ayala maintains that he still did not believe that Volkswagen was doing anything criminal. But he was bothered enough by the results of the WVU research to launch a compliance project. It would be a much more intense inquiry than WVU had done, one that would take advantage of CARB's clout as a regulator. Ayala assembled a team of specialists and began collecting a small fleet of Volkswagens to test, borrowing the cars from VW owners who were paid a fee and received temporary replacement vehicles. Soon a major investigation was underway. In line with CARB confidentiality rules, WVU and ICCT were kept out of the loop. Ayala did not even tell Mary Nichols, the chairwoman of CARB. She is a political appointee and was not informed about the Volkswagen case in order to avoid any suggestion of political influence. Carder and his team had no inkling of the forces they had set in motion. "It was interesting," Kappanna said of the test data. "But we never thought it would blow the lid off Volkswagen. Never, ever."

CHAPTER 15

Exposure

BERND GOTTWEIS, Volkswagen's head of product safety, was known inside the company as the Red Adair of quality control. Adair was a Texan famous for his ability to extinguish runaway oil well fires. Gottweis, a Volkswagen veteran who reported directly to Winterkorn, also had a talent for putting out fires, in his case the kind that occurred when there was a serious defect in a Volkswagen product. Soon after the West Virginia University researchers reported their results, Gottweis was among the first to get a copy. In May 2014, after taking stock of the situation, Gottweis wrote a one-page report that was included in the packet Martin Winterkorn's aides put together for him to read over the weekend. The packet was known internally as "Wikopost" ("Wiko" was one of Winterkorn's nicknames). It contained some unsettling news.

One of the Volkswagens tested by the team from West Virginia had emitted fifteen to thirty-five times the permitted amounts of nitrogen oxides during road tests, Gottweis wrote. The other, the Passat with the urea-based SCR system, was five to eighteen times over the limit. The West Virginia team had also tested a BMW as part of its work for the International Council on Clean Transportation, Gottweis noted. The BMW had not shown a discrepancy between emissions on the road and in the testing lab, except when the car was driving up a mountain and the engine was under particular strain. The reason, although Gottweis did not say so,

Ferdinand Porsche (left, pointing) shows Hitler and his minions a model of the "people's car" in 1938. It was largely a propaganda exercise; only a handful were built before war broke out. *(Heinrich Hoffmann, Hulton Archive, Getty Images)*

Machinery inside the factory in 1944. Though designed to produce the Beetle, the Volkswagenwerk shifted almost exclusively to production of military equipment during the war. *(Volkswagen Aktiengesellschaft)*

Allied raids on the Volkswagen factory caused roofs to cave in and walls to collapse, as in this photo from 1944, but most of the machinery survived, and production was only temporarily interrupted. *(Volkswagen Aktiengesellschaft)*

Ivan Hirst (in beret) was a resourceful British major assigned to oversee Volkswagen after the German surrender. Shown here in 1946, Hirst revived production of the Beetle and was effectively the company's first postwar chief executive. *(Volkswagen Aktiengesellschaft)*

Heinrich Nordhoff, a German appointed chief executive in 1948, was a capable manager but also an authoritarian who held on to his post for two decades. *(Volkswagen Aktiengesellschaft)*

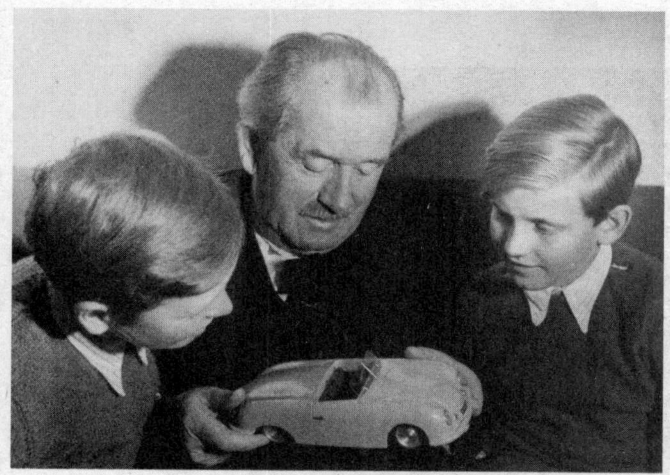

Ferdinand Porsche with his grandsons Ferdinand Alexander Porsche (left) and Ferdinand Piëch in 1949. Piëch's impact on the auto industry may well have turned out to be greater than that of his grandfather. *(Historical Archive at Porsche AG)*

The Beetle and its cousin, the Porsche 356, became unlikely symbols of 1960s rebellion. The singer Janis Joplin, shown here in 1969, had her Porsche custom painted. *(Sam Falk, The New York Times Photo Archives 1969)*

The New Beetle, launched in 1997, tapped into nostalgia for the original Beetle and was a hit in the United States. The model shown here is from 2017. *(Andreas Rentz, Getty Images)*

A view of the main factory in Wolfsburg on the banks of the Mittelland Canal in 2007. In the foreground is the entrance to the Autostadt, a museum and temple of auto worship. *(Collection ullstein bild, Getty Images)*

Ferdinand Piëch at the wheel of an Audi at Volkswagen's annual shareholders meeting in Hamburg in 2010. Like Steve Jobs of Apple, he was one of the few chief executives whose personal stamp was visible on their company's products. *(David Hecker, DDP, Getty Images)*

Ursula Piëch, the former family governess better known as Uschi, was one of the few people whom Ferdinand Piëch trusted. He arranged for her to become a member of the Volkswagen supervisory board. They attended an auto industry awards ceremony in Berlin in 2011. *(Sean Gallup, Getty Images Entertainment)*

Martin Winterkorn, the Volkswagen chief executive, shown at the Geneva International Motor Show in March 2015, was known for his obsession with detail and for his temper. *(Fabrice Coffrini, AFP, Getty Images)*

Wendelin Wiedeking, former chief executive of Porsche (left), and Holger Härter, the sports car maker's former chief financial officer, at a Stuttgart court in 2015 during their trial on charges of stock market manipulation. Both men were acquitted. *(Thomas Kienzle, AFP, Getty Images)*

Wolfgang Porsche and his partner, Claudia Hübner, at the Vienna Opera Ball in 2016. After his cousin Ferdinand Piëch lost a power struggle in 2015, Wolfgang Porsche became the most powerful member of the families on the Volkswagen supervisory board. *(Gisela Schober, German Select, Getty Images)*

Hans Dieter Pötsch (left), the chairman of the supervisory board, and Matthias Müller, the chief executive, were both longtime insiders whose appointment in the wake of the emissions scandal raised questions about whether Volkswagen would change its ways. They are shown at a news conference in December 2015. *(Carsten Koall, Getty Images News)*

Francisco Javier Garcia Sanz (left), a Spaniard, and Volkswagen's longtime procurement chief, handled settlement negotiations in the United States. He is shown in 2016 with Matthias Müller, the Volkswagen chief executive, and Christine Hohmann-Dennhardt, a former judge appointed to strengthen Volkswagen's internal compliance system. *(John McDougall, AFP, Getty Images)*

Mary Nichols, the chairwoman of the California Air Resources Board, which did much of the detective work that exposed Volkswagen's emissions fraud (2008 photo). *(Irfan Khan, Getty Images)*

Dan Carder, who oversaw the study by West Virginia University that first raised questions about Volkswagen emissions, was named one of *Time* magazine's 100 most influential people of 2015, along with Vladimir Putin and Pope Francis. He attended a *Time* gala in New York with his wife, Hillarey, in 2016. *(Taylor Hill, FilmMagic, Getty Images)*

Customers can take delivery of their cars at the Autostadt. By 2016, when this photo was taken, Volkswagen was losing market share in Europe. *(Sean Gallup, Getty Images News)*

was that BMW had not scrimped on the antipollution technology it used in cars sold in the United States. Whereas the Volkswagens had either a lean NOx trap or an SCR system, the BMWs had both.

CARB had already declared its intention to do further tests, Gottweis wrote. The California regulator had also asked Volkswagen for technical details about the pollution control equipment, including how the engine software regulated doses of the urea solution.

Gottweis was blunt about the bind that the testing had created for Volkswagen. "A thorough explanation for the dramatic increase in NOx emissions cannot be given to the authorities," he wrote. "It can be assumed that the authorities will then investigate the VW systems to determine whether Volkswagen implemented a test detection system in the engine control unit software (so-called defeat device) and, in the event a 'treadmill test' is detected, a regeneration or dosing strategy is implemented that differs from real driving conditions.

"In Drivetrain Development, modified software versions are currently being developed which can reduce the real driving emissions," Gottweis continued, "but this will not bring about compliance with the limits, either. We will inform you about the further development and discussion with the authorities."

Gottweis's memo delivered a clear warning to the very highest level of management that Volkswagen had been caught using a defeat device, and that it had no excuses and could not easily rectify the excess emissions. Although Volkswagen does not dispute that Winterkorn received the Gottweis memo, Volkswagen has argued that there was no proof that Winterkorn actually read the memo, which was included in a stack of other documents. Even if Winterkorn did read it, according to Volkswagen, he might have concluded that his subordinates were taking care of the problem. It is hard to believe Winterkorn would have reacted so calmly. A memo from the company's star troubleshooter was not something to ignore. But even if Winterkorn did not read the memo or appreciate the severity of the problem, the issue came up again that same month during a high-level meeting in Wolfsburg with representatives of Bosch, the company based in Stuttgart that supplied the computers and software used to manage the Volkswagen diesel motor. The purpose of the May 28 meeting was to discuss the partnership between Volkswagen and Bosch in the United States. According to a document that has been variously identified, in published reports, as a meeting agenda and as notes of the same meeting, the attendees included not only Winterkorn but also Volkmar Denner, the chief executive of Bosch. One of the items on the agenda was the "acoustic function," the code name for the defeat device. Bosch maintains that the meeting included a general discussion about diesel engine acoustics, and was not about a defeat device.

If Volkswagen had been honest with the EPA and CARB at that point, and admitted to officials that its diesels in the United States were programmed to foil emissions regulations, the damage to the company's reputation and finances would very likely have been serious but not as devastating as it turned out to be. Similar cases suggest that Volkswagen would have paid a fine in the hundreds of millions of dollars or perhaps only tens of millions. It would have reached an agreement with the EPA to recall the cars to make them as compliant as possible with clean air rules, probably without explicitly admitting wrongdoing. The public would not have learned about the problem until the settlement and recall had been announced, as is customary in such cases. After a few days of negative headlines, the media would in all likelihood have moved on.

But Volkswagen did not exploit the chance to be transparent. That was not the way people in Wolfsburg ticked. They had already passed up several opportunities to make Volkswagen diesel vehicles legal. They had received plenty of warnings that something was amiss. Instead, in their zeal to meet Winterkorn's and Piëch's astronomical sales targets, they had expanded the use of defeat devices to include millions of cars. In the years preceding the West Virginia University study, Volkswagen had also considered detailed proposals, made on several occasions at product strategy meetings, for improving the pollution control equipment so that emissions under normal driving conditions would better match those achieved in the lab. The proposals were rejected. Too expensive, Winterkorn and others said. The use of defeat devices, which may have begun as a stopgap, had become a habit and, as long as the deception remained undiscovered, a cost advantage. Now, in 2014, Volkswagen gambled against the odds that it could continue to avoid the consequences.

An internal Volkswagen presentation, prepared soon after executives learned of the West Virginia tests, discussed various strategies the company could adopt to address regulators' suspicions. The tone of the presentation was dispassionate, a cool assessment of risks and costs. One option was for Volkswagen to simply refuse to acknowledge there was a problem—in other words, to stonewall, or lie again. Another option was to offer to update the engine software. But the update would not bring emissions down to the required levels, the presentation said, coming to the same conclusion as Gottweis. In the worst case, Volkswagen could admit there was a problem and buy back diesel cars sold in the United States. The last option does not appear to have been seriously considered at the time.

Despite Volkswagen's claim to be a pioneer in sustainable transportation, no documents have emerged in which Winterkorn or anyone else demanded an internal investigation into the excess emissions. There were no calls for disciplinary action against those responsible, no sense of shock or outrage. The documents available, including those Volkswagen later filed in its defense, convey the impression that the emissions issue was a technical problem and a regulatory headache, not an occasion for soul-searching. No one inside Volkswagen seemed terribly surprised to learn that company products may have contained a defeat device. No one, as far as is known, saw the revelations as a reason for the company to examine whether the wrongdoing was a symptom of deeper corporate dysfunction. On the contrary, Volkswagen executives coolly weighed how much they should reveal to regulators. "It should first be decided whether we are honest," Oliver Schmidt, Volkswagen's top compliance official in the United States, wrote to a colleague.

The West Virginia report, which the International Council for Clean Transportation posted on its website without fanfare in May 2014, came at an especially awkward time for Volkswagen. The company was in the midst of applying for regulatory certification of a new generation of diesels for the 2015 model year. The Passats, Golfs, Jettas, and others came equipped with an updated version of the EA 189 motor, designated the EA 288, which was supposed to meet a new, tougher standard for nitrogen oxide emissions that took effect in 2015. All of the Volkswagen vehicles sold in the United States would now be equipped with the system that used squirts of the AdBlue urea solution to neutralize nitrogen oxide emissions in the exhaust. Previously only the Passat and the larger Audis and Porsches with three-liter diesel motors had the equipment. It was the same so-called SCR technology Volkswagen had considered and rejected during the short-lived BlueTec alliance with Daimler years earlier.

But Volkswagen continued to believe that Americans would not buy a car that required them to refill a tank with an unfamiliar chemical. So the cars were programmed to ration doses of the diesel exhaust fluid, except when the computer that controlled the engine detected the telltale signs of an official test. In short, the latest models were also programmed to cheat.

As regulators began pressing Volkswagen to explain the findings by West Virginia University, the company realized it had to do more to reduce emissions under normal driving conditions. In June, Volkswagen quietly tweaked its application for EPA certification to say that the urea tank would need to be refilled "approximately" every ten thousand miles. The addition of that one word was significant. Volkswagen was no longer promising owners they would be able to go between oil changes without topping up the urea tank. The company was tacitly admitting that the existing emissions system used too little of the urea solution to be effective. As new Tiguan SUVs and other German-made cars were already piling up at U.S. ports, waiting for regulatory approval to go on sale, Volkswagen made plans to update the software in the 2015 models to increase the dosage of the urea solution in order to better neutralize nitrogen oxides.

Among themselves, Volkswagen executives were anxious about the decision. They fretted that asking owners to perform an extra chore would present a "significant rejection reason to potential buyers." According to estimates by Volkswagen engineers, consumption of the urea solution would nearly double from 0.8 liters every 1,000 miles to 1.5 liters. Many owners would have to refill the tanks as soon as 6,000 miles, depending on how aggressively they drove and what kind of routes they traveled. But even with the increased urea consumption, Volkswagen was aware that reprogramming the engine computer would not be enough to bring nitrogen oxide emissions to within the legal limit.

Volkswagen could ill afford anything that would turn off buyers. Diesel sales had climbed rapidly starting in 2009, peaking at 111,000 in 2013. But by 2014 Volkswagen TDI sales had begun to slip, as had sales of cars with conventional gasoline motors, which still accounted for more than three-quarters of Volkswagen's business in the United States. The New Beetle was no longer a novelty, and initial enthusiasm for the Chattanooga-made Passat was wearing thin. The coming 2015 models, including a new generation of the Golf, offered a chance to revive the growth. Any delay would be catastrophic.

Meanwhile, Alberto Ayala and his team at CARB were pressing the company politely but persistently to explain the emissions anomalies. CARB did its own road testing by means of portable emissions-measuring equipment, as well as stationary tests in the El Monte lab. But though the tests showed that something was wrong, they did not explain why. At that point, CARB officials say, they were not trying to expose any wrongdoing by Volkswagen. They were just trying to figure out what the problem was so that it could be fixed.

It was a slow process. CARB would present the results of its testing. Volkswagen, which had a technical facility in Oxnard, west of Los Angeles, would perform its own tests. Then engineers from CARB and Volkswagen would meet to discuss their findings. They often gathered in an unadorned conference room equipped with a veneer conference table, gray carpet, and brown swivel chairs at CARB offices adjacent to the El Monte lab. The meetings were highly technical and lasted hours, sometimes all day. A nonspecialist listening in would not have understood a word. "For every answer we got, we generated a couple more questions," Ayala said. CARB took the lead in dealings with Volkswagen, but Ayala also kept the EPA informed.

AS THE BACK-AND-FORTH dragged on without a solution, Ayala grew impatient. The Volkswagen executives responsible for dealing with regulators gave answers that the regulators regarded as evasive, nonsensical, or dismissive. CARB's testing was wrong, Volkswagen complained. The outside air pressure threw off the results. The routes followed were inconsistent.

Efforts to figure out what was amiss with the Volkswagens began consuming so much of the CARB technicians' time, as well as space in the four test bays in El Monte, that it was crowding out other essential work. Volkswagen was slowing down the testing needed to certify vehicles from other manufacturers. Ayala was also concerned that some seventy thousand Volkswagen diesels were on California roads, polluting much more than allowed. CARB could be accused of not doing its job if it didn't find a solution.

To try to resolve the engineering stalemate, officials from both sides held a conference call on October 1, 2014. The Volkswagen representatives included Schmidt, a German who was head of the company's Environmental and Engineering Office in Auburn Hills, Michigan; and Stuart Johnson, an American who was his second in command. (Johnson was among the U.S. employees who had informed people in Wolfsburg back in 2006 about the legal risks of using defeat devices.) While repeating explanations that officials considered implausible, the Volkswagen executives unveiled their plan to perform a recall in order to update the engine software for diesels starting with the 2009 model year. The update would "optimize" performance of the emissions equipment in all of the "clean diesel" vehicles that the company had sold to date, Volkswagen said.

The planned recall seemed like a concession by Volkswagen, but was more likely simply a delaying tactic. The company still did not provide an honest explanation for the excess emissions. Instead, it provided assurances to the regulators, which proved to be false, that the recall would cure the excess emissions exposed by the West Virginia study. Volkswagen told CARB officials, "The new software incorporates the latest engineering experiences to enhance the efficiency of the SCR system." For older cars with a lean NOx trap and no urea tank, the so-called software upgrade would reduce "the frequency of unnecessary replacement of after treatment system components," Volkswagen told the EPA. The company added that the vehicle's engine management software would be modified to improve operation of the soot particle filter "under extreme driving conditions." These answers were, at best, incomplete.

Volkswagen was similarly misleading in the explanation it gave to dealers and customers for the recall, which was carried out in early 2015. It claimed that the new software installed during the recall was needed to prevent a malfunction lamp from illuminating. The recall was also "part of Volkswagen's ongoing commitment to our environment." Volkswagen did not mention that owners would need to refill the urea tank more often and that fuel economy would suffer in their cars. EPA and CARB officials, who still believed they were dealing with a technical issue and not deliberate wrongdoing, approved the recall. Accepting Volkswagen's assurances that the recall would fix excess emissions, they allowed the 2015 diesels to go on sale.

Volkswagen brought cars in for the software fix, known as a "flash action," in early 2015, eventually updating the software in 280,000 vehicles. Afterward, the cars polluted less than they had, but the upgrade did not remove the illegal code. In fact, Volkswagen brazenly used the recall to enhance the effectiveness of the defeat device. The flash action programmed the cars to go into good-behavior mode when the wheels were moving but the steering wheel remained stationary, as would be the case during a lab test on rollers. The change was another way of ensuring that the cars would recognize when they were under scrutiny, and generate exemplary emissions readings.

Meanwhile, Audi was facing questions from CARB and the EPA about its three-liter diesel motor, which was also used in the Porsche Cayenne SUV and the Volkswagen Touareg SUV. The regulators were curious whether the three-liter engines in the larger SUVs had the same excess emissions as the Volkswagen two-liter motors. In fact, Audi had done its own road testing starting in late 2014 and found that nitrogen oxide emissions in some vehicles were as much as ten times the legal limit. Audi did not fully disclose those results to regulators, though. It merely told them about tests of a diesel A8, a luxury sedan, in which emissions were three times the limit. Audi engineers provided explanations for those high emissions that the regulators considered evasive or spurious. At one point, Audi representatives attributed high emissions found by CARB to the unique driving conditions in Los Angeles and insisted that the pollution control equipment was the best it could be, given that it had to fit under the floor of the vehicle. "We heard all kinds of explanations for the discrepancies in emissions that we were finding," Ayala said. "This is one example of how they would often try to skirt the issue by throwing a few fancy technical terms into their answer."

After the Volkswagen recall was complete, CARB officials asked for data on whether the software update had brought emissions down to within legal limits. When Volkswagen representatives did not provide the information—which would have shown that the recalled vehicles hadn't fixed the excess pollution—CARB officials asked again. And again. Receiving answers they felt were unclear, CARB informed Volkswagen that it would do another round of testing.

This time the tests would be even tougher. CARB planned to put the cars on rollers in a lab, but the procedure would last longer than usual and follow different simulated driving patterns. Such testing would confuse any defeat devices, which were programmed to kick in when they recognized the standard driving pattern. The defeat devices might not work, and CARB officials would discover how much the cars really polluted. The CARB technicians also plugged into the cars' computers, known as onboard diagnostic units or OBDs. If the OBDs had been functioning properly, they would have signaled malfunctions in the emissions equipment. Yet the OBDs indicated that the systems were working. Dashboard warning lamps that would have indicated a problem did not go on. That made no sense.

On May 18, 2015, a Volkswagen engineer wrote an e-mail to Stuart Johnson and several managers in the Volkswagen department that developed engines. The engineer expressed concern about what the latest round of testing would show and asked, "Do we need to discuss next steps?" He also wanted to know how he should respond to questions from CARB about performance of the filter that removed soot from the exhaust. "Come up with the story please!" he wrote.

A few days later, another executive warned that CARB was subjecting Volkswagen to an unusual level of scrutiny. "Please be aware that this type of action from California ARB staff/management is not a normal process," wrote Michael Hennard, senior manager of emissions compliance. "We are concerned that there may be possible future problems/risks involved. It should also be noted that this TDI software issue is being reviewed and monitored by upper management at CARB." According to the various state complaints, the reaction inside Volkswagen was an admonishment from a senior manager not to be so frank in an e-mail. Exposure was now inevitable. Yet the cover-up continued.

CHAPTER 16

Piëch's Fall

THE MANEUVERING BETWEEN Volkswagen and regulators in the United States was invisible to most of the world. A few industry insiders had heard rumors that Volkswagen had some kind of problem in the United States. But nothing leaked out to the media or the general public. There was plenty of attention on Volkswagen in early 2015, but it was due to a completely different drama. As so often, the chief protagonist was Ferdinand Piëch.

In April 2015, Piëch used the German magazine *Der Spiegel* to signal that a Volkswagen manager had fallen out of favor. But this time it wasn't just any manager. "Ich bin auf Distanz zu Winterkorn," Piëch told *Der Spiegel*. "I'm distancing myself from Winterkorn."

It was classic Piëch to render a manager's position untenable by subtly but unmistakably undermining the man in public. Piëch once signaled his displeasure with Wendelin Wiedeking by telling reporters during a test drive junket that the Porsche chief executive "was the best" man for the job. The reporters immediately picked up on the nuance: Piëch didn't say Wiedeking "*is* the best." He said he "*was* the best," speaking of Wiedeking in the past tense. Wiedeking was soon gone. But Winterkorn was different. He and Piëch had been two sides of the same coin for decades. Piëch was the strategist; Winterkorn was the executive who made it happen. Winterkorn called Piëch his

Ziehvater, a word that can mean "mentor" but also "surrogate father." It had become hard to imagine one without the other, and unthinkable that they would be on opposite sides of a power struggle.

This time, though, Piëch seemed to have picked the wrong fight. Winterkorn, more confident after seven years as chief executive, refused to capitulate to Piëch's attempt to undermine him publicly. Other members of the supervisory board were annoyed that Piëch was trying to push out Winterkorn without consulting them. The state of Lower Saxony, with its 20 percent of the voting shares, backed Winterkorn. Crucially, so did the Volkswagen workers council and its powerful chairman, Bernd Osterloh, who functioned as a kind of shadow supervisory board chairman. Osterloh issued a statement calling Winterkorn "the most successful automobile manager" and saying his contract should be extended. Together, the state and the labor representatives had a majority on the supervisory board. Piëch had no prayer of mustering the two-thirds majority required to fire Winterkorn. At seventy-eight, Piëch seemed to have lost his Machiavellian gifts.

On April 25, 2015, less than two weeks after Piëch had provoked the controversy, he offered his resignation to the supervisory board, as did his wife, Ursula. Piëch may have expected the rest of the supervisory board to refuse his resignation. If so, he was wrong. After more than two decades as the company's dominant figure, Piëch no longer held any official position at Volkswagen.

Ironically, Piëch's exit came only weeks before Volkswagen achieved its goal of becoming the largest carmaker in the world. Figures from the second quarter of 2015, published in July, showed that Volkswagen had surpassed Toyota in the number of cars sold worldwide. Volkswagen was on track to sell ten million cars in the course of the year, hitting its sales target two years ahead of schedule.

Curiously, though, Volkswagen let the milestone pass without even a self-congratulatory press release. Celebration was probably premature. The margin over Toyota was slim, about 200,000 cars. A bad quarter would be enough to push Volkswagen back into second place. And Volkswagen was still far from reaching the other targets that were part of Strategy 2018. The company's profit margin in the first half was about 6 percent, short of the goal of 8 percent, and far behind Toyota's margin of 10 percent. Volkswagen continued to be much less efficient than its Japanese rival. Even without the brewing disaster in the United States, which was not yet common knowledge, it was too early to set off fireworks in Wolfsburg.

What motivated Piëch to break with Winterkorn remains a mystery. One theory was that Wolfgang Porsche, Piëch's cousin and rival, had succeeded in drawing Winterkorn into his camp. Though the Piëch and Porsche clans voted as a bloc in the Volkswagen annual meeting, behind closed doors the antagonism between the two sides of the family remained intense. Some Volkswagen observers detected signs that the relationship between Winterkorn and Wolfgang Porsche had become chummier, perhaps too chummy for Piëch's liking. Winterkorn's increasing independence posed a threat to Piëch's dominance not only of the company but also of the Porsche and the Piëch families. Eventually, Piëch would have felt pressure to cede chairmanship of the supervisory board to Winterkorn, when it was time for Winterkorn to retire as operational chief of Volkswagen.

Another theory was that Piëch foresaw the turmoil that the emissions problem would cause, and engineered an excuse to exit. According to this scenario, Piëch provoked the power struggle knowing he would lose. It was indeed surprising how quickly Piëch resigned as supervisory board chairman once it became clear that his attempt to defenestrate Winterkorn would fail. In other respects, the theory seems far-fetched. Piëch was a fighter and Volkswagen was his lifework. It is hard to believe he would head for the lifeboats at the first sign of an iceberg.

The most plausible explanation may be the most straightforward one. Piëch had come to believe that Winterkorn was not doing a very good job. Despite the investment in the Chattanooga factory and the clean diesel marketing push, Volkswagen was far from its goal of selling 1 million cars in the United States. First-half sales indicated that Volkswagen would be lucky to sell 600,000 cars in the United States during the year, even after adding Audi and Porsche to the figures. The

Volkswagen brand's market share in the United States was only 2 percent, and was declining. If Piëch was aware of the emissions problem by March 2015, as German media have reported, the issue may have added to his discontent with Winterkorn's leadership.

Globally, Volkswagen was dangerously dependent for profit on Audi and Porsche, its luxury brands. Cars with the Volkswagen badge accounted for half of sales revenue, but made hardly any money because manufacturing costs were too high. The operating profit margin for the Volkswagen passenger cars division, weighed down by the untouchable German workforce, was less than 3 percent. Investors were losing faith in the company's long-term prospects, and the company's share price was falling. Some in the German news media speculated that Piëch wanted to replace Winterkorn with a new favorite, Matthias Müller, the chief executive of the Porsche division, which was thriving.

Whatever the truth, Piëch's exit may have been fortunate from his point of view. Though he no longer was chairman, he retained influence behind the scenes. He owned 14 percent of Porsche Automobil Holding, the vessel for the Porsche-Piëch majority stake in Volkswagen. He was a member of the holding company's supervisory board and the most prominent member of the Piëch family shareholders. He remained a member of the Porsche shareholder's committee, a secretive panel that exerted powerful influence behind the scenes over strategy and decision making. When Piëch returned to Wolfsburg in September 2015 for a meeting of the shareholder's committee, arriving in a red Bentley driven by Ursula, his mere appearance sparked speculation that he was plotting a comeback. So formidable was Piëch's reputation as a corporate power player that no one would believe that he was really gone until he lay in his grave, and perhaps then only if he had a stake through his heart.

Piëch's resignation from the supervisory board in April 2015 meant he was out of the line of fire by the time the world learned of the emissions fraud the following September. There were news stories and commentaries blaming Piëch for creating the company culture that nurtured the deception. But he did not have to take responsibility for crisis management following the EPA disclosures or endure debate about whether he should suffer consequences. He was already gone.

Berthold Huber, a former president of the IG Metall labor union who was deputy chairman of the supervisory board, was elected acting chairman to replace Piëch until a successor could be chosen. It was another manifestation of organized labor's extraordinary influence over Volkswagen management. Among some members of the supervisory board, there was an undercurrent of relief that they would no longer have to endure Piëch's machinations. "It is no exaggeration to say that he is one of the most important people in the history of German business," said Stephan Weil, the prime minister of Lower Saxony. "Nevertheless," the center-left Social Democrat added, "it was urgently necessary to end the speculation about persons and to ensure clarity in top management. Volkswagen and its many thousand employees must be able to concentrate on business."

Still, Piëch's departure left a vacuum. He had led Volkswagen from near bankruptcy in the early 1990s to the pinnacle of the auto industry. Like Steve Jobs at Apple, Piëch was among a handful of executives who put an unmistakable personal stamp on the companies they ran. Like Jobs, Piech was obsessive about the details of product design and had a genius for creating objects of desire. Under his supervision, Audi grew from a maker of dull middle-class sedans to a luxury carmaker on a par with BMW and Mercedes. When Piëch took over in Wolfsburg, the Volkswagen brand was still struggling to find its post-Beetle identity. Piëch redefined the idea of the people's car to mean German engineering for the masses. With his focus on the minutiae of manufacturing, Piëch improved the workmanship of Volkswagens and made them feel like more expensive cars. Almost everywhere except the United States, the strategy was incredibly successful. In Europe, Volkswagen and its related brands had a market share of 26 percent in April 2015, much more than double that of the nearest competitor, PSA Peugeot-Citroën of France.

But Piëch also left behind enormous problems. He appeared indifferent to outside shareholders and seemed more focused on empire building than on making money. He owed his power on the supervisory board to a de facto pact with labor not to impose major job cuts. (The moment Osterloh backed Winterkorn, signaling that Piëch had lost labor support, his power evaporated.) Volkswagen remained inefficient, and marginally profitable compared with its most important competitors. He had a weakness for costly prestige projects, like the ill-fated Volkswagen Phaeton and its glass-walled factory in Dresden, or the Bugatti roadsters that cost more than a million dollars but still lost money.

Piëch created a culture inside Volkswagen that revolved around him and a few protégés. Even the most loyal of them, like Winterkorn, could be jettisoned at any time. Centralized authority created decision-making bottlenecks. Winterkorn perpetuated the system. The rest of the organization was paralyzed until a ruling came down from the top. The engineers dominated, which helped create excellent products but also meant that other areas like compliance were neglected. During Piëch's tenure, the company had been involved in behavior that was outright criminal, like the López affair or the prostitution scandal. Underlings always suffered the consequences. Any reforms were weak and, as subsequent events proved, inadequate. It was a toss-up what history would remember most about Piëch: the great cars or the scandals.

CHAPTER 17

Confession

IN JUNE 2015, after Volkswagen finished recalling its diesel fleet in the United States, CARB tested a 2012 Passat, which was equipped with the urea emissions system. Not surprisingly, the tests showed that the recall and software update had not solved the problem of excess pollution. Output of nitrogen oxides rose after about twenty-three minutes of driving, CARB said, one minute after the end of the standard test cycle. During some simulated driving cycles, CARB found, there were no injections of the urea solution at all. CARB demanded that Volkswagen show it the software code that controlled the urea dosing in the new 2016 models, which were already rolling off assembly lines. CARB also wanted to see the code in the older models, so it could do a before-and-after comparison and check whether, as Volkswagen claimed, the emissions problems had been fixed in the new cars. The idea that CARB would take a closer look at the older models raised even more alarm inside Volkswagen. "We must be sure to prevent the authority from testing the Gen 1!" one employee wrote in an e-mail, referring to the first generation of clean diesels. "If the Gen 1 goes on to the roller at CARB, then we'll have nothing to laugh about!!!!!" CARB then made a threat. If Volkswagen failed to comply with its request, CARB said, the agency would refuse to approve the 2016 models for sale in California.

In the regulatory world, threatening to withhold approval is the nuclear option. Despite stereotypes about regulators as overzealous bureaucrats—real-life versions of the Green Police in the 2010 Audi commercial—they tend to be cautious and methodical. If a company cannot get approval to sell its cars in California, it can't sell them anywhere in the United States. A dozen other states, including New York, Pennsylvania, Maryland, and most of New England's, follow California's clean air standards and do not approve any car rejected by CARB. No vehicle can be a success in America if it is banned from so many states. The EPA, too, does not certify cars that have been rejected by California. As a result, assembly lines may come to a halt. Jobs are endangered. The costs to the carmaker are enormous, to say nothing of the damage to its reputation. Regulators are reluctant to impose those kinds of consequences unless they are sure of their facts and the violation is severe.

For Volkswagen, the lack of certification to sell the new cars was creating a critical situation. By mid-July, the 2016 Volkswagen models had begun piling up in ports, stuck there until the emissions problem was clarified. Michael Horn, the German president of Volkswagen of America—largely a marketing job—sounded the alarm. Horn notified numerous top managers in Germany of the danger that the 2016 models would not be approved unless Volkswagen provided enough information to satisfy CARB. Among those Horn supposedly informed were Christian Klingler, a member of the Volkswagen management board in charge of sales and marketing, and Heinz-Jakob Neusser, head of technical development for Volkswagen brand cars. They were among the most important people in the company, and both dealt directly with Winterkorn. (Volkswagen lawyers argued there is no proof that Horn was aware that the emissions problem was the result of a defeat device.)

In Wolfsburg, CARB's threat to withhold approval for the 2016 vehicles was a topic at a regular meeting on July 21, 2015, of an internal Volkswagen committee that discussed safety and regulatory issues. The committee's conclusions were summarized in a memo to Frank Tuch, the chief of quality control. Item 6.3 of the memo, which also dealt with other issues, was headed "Diesel Motors USA." It noted that certification of the 2016 models had stalled because questions about excess emissions of earlier models had not yet been answered. The committee decided to create a task force headed by Friedrich Eichler, the company's head of engine development. The group's goal would be to achieve "fast and effective de-escalation of the issue with officials." Volkswagen should approach the regulators "offensively," the memo said.

On July 27, 2015, Winterkorn was briefed by engineers on the diesel problem in a small group that also included Herbert Diess, a former BMW executive who had become head of the Volkswagen brand passenger cars at the beginning of the month. In court documents, Volkswagen lawyers maintained that Winterkorn had by then still not registered the gravity of the problem. That is implausible, given Winterkorn's attention to detail and the amount of information he had already been provided about the issue. But even according to Volkswagen's version of events, Winterkorn and Diess were in all likelihood informed that the underlying problem related to "software altered to influence emissions performance during testing," according to Volkswagen's account of the meeting. But according to Volkswagen, Winterkorn was not informed at the meeting that the software violated United States law. Yet Winterkorn still kept shareholders and the supervisory board in the dark. (Winterkorn later said under oath that he did not hear the term "defeat device" until September 2015.)

VOLKSWAGEN BEGAN TO BELATEDLY face up to the likelihood that the emissions problem would have legal consequences. One of Volkswagen's in-house lawyers in Wolfsburg, Cornelius Renken, who was responsible for product safety, had been informed by subordinates no later than May 2015 about the possibility that the emissions irregularities were linked to the engine software. In late July, Volkswagen asked the Washington office of Kirkland & Ellis, a law firm with expertise in regulatory issues, to provide an opinion on the potential penalties. In early August, the firm issued a five-page, single-spaced analysis, addressed to Renken and David Geanacopoulos, a lawyer who was Volkswagen of America's senior executive vice president for public affairs and public policy. The memo examined what would happen if Volkswagen was caught deploying a defeat device. The opinion noted that the EPA had since the mid-1990s "been aggressive in efforts to deter the use of vehicle engine or emissions control software designs or features that deactivate, or make less effective, a vehicle's emissions control system" outside test conditions. That was another way of saying that regulators could be expected to take a hard line if they found a defeat device. "It is unlikely," the Kirkland & Ellis memo stated, that Volkswagen would be able to avoid paying civil penalties to the government.

But the opinion provided some reassurance to Volkswagen about what the financial impact might be. Noting that virtually all such cases were settled out of court, Kirkland & Ellis reported that the largest fine to date was $100 million imposed the previous November on Hyundai-Kia. The Korean carmaker had admitted overstating fuel economy and understating greenhouse gas emissions of 1.1 million cars and light trucks sold in the United States, double the number of Volkswagens under suspicion. That was a lot of money, to be sure, but not catastrophic for a company like Volkswagen with more than €200 billion, or about $240 billion, in annual sales. The opinion also raised the possibility that Volkswagen would face class action suits from owners if, for example, repairs to the cars' emissions systems led to lower fuel economy.

The Kirkland & Ellis opinion created the impression that the potential consequences for Volkswagen were serious but manageable. That was the message that Volkswagen executives took away. They still believed they could somehow work out a settlement with CARB and the EPA. But Volkswagen overlooked the memo's point that cooperation is a factor in such settlement negotiations. In the past, carmakers had almost always acquiesced immediately when confronted with evidence of emissions violations, and cooperated with the EPA. That was true of Hyundai-Kia, whose cooperation was cited as a factor in the settlement it reached with the EPA and CARB.

Volkswagen had taken a far different approach. By August 2015, the company had spent more than a year procrastinating, providing regulators with false, misleading, or incomplete information. It had carried out a recall that did not deliver the promised improvements and appeared to be simply a ruse designed to buy time. The company had continued to sell cars with illegal software, including the 2015 model year, despite clear notice from regulators that there was a serious problem. Not least, it had failed to provide details about its engine software despite repeated requests from CARB.

The cover-up arguably went back to the very beginnings of the defeat device. Since 2008, Volkswagen had sworn the standard EPA oath, which included a promise not to sell cars that "cause the emission into the ambient air of pollutants in the operation of its motor vehicles or motor vehicle engines which cause or contribute to an unreasonable risk to public health or welfare." Some Volkswagen employees might be able to argue that they were ignorant of the illegal software. But at least one of the people dealing with regulators during the clean diesel drive also helped contrive the original defeat device. James Robert Liang, a Volkswagen employee since 1983, was part of the team in Wolfsburg that designed the EA 189 motor. He later admitted having been among those who designed the cycle-beating software when it became clear the motor could not meet American emissions standards. After transferring to the United States in 2008, Liang worked at Volkswagen's testing center near Los Angeles and was among those shepherding cars through the certification process. He and others repeatedly fed the regulators with false information. After the West Virginia study came out, Liang was in the group desperately trying to keep CARB and the EPA from finding out about the defeat device. Volkswagen could not expect the regulators to be magnanimous when they discovered the truth.

Volkswagen nevertheless continued to obfuscate. In early August, Oliver Schmidt, who the previous March had moved to a job in engine development in Wolfsburg, and Stuart Johnson, who had replaced him as head of Volkswagen's Environmental and Engineering Office, admitted to CARB that the recall early in the year had not fixed the emissions problems in the cars then on the road. But they continued to conceal the existence of a defeat device. On August 5, Schmidt and Johnson asked to meet with CARB's Ayala after learning he was scheduled to speak at an industry conference in Traverse City, Michigan, organized by the Center for Automotive Research, a nonprofit group. According to Ayala, Volkswagen booked a meeting room at the conference venue, a resort on the shores of Lake Michigan. Schmidt and Johnson arrived at the appointed time with a thick binder of technical information.

They spent two hours going over the information with Ayala. To Ayala, the information looked credible at first glance. Despite everything, he still trusted the Volkswagen executives. In fact, Ayala was elated that Volkswagen finally seemed to have cracked the problem. After he returned to California, Ayala turned the binder over to his staff. A week later, the compliance engineers came back to him with the results of their analysis. The information provided by Volkswagen was all nonsense. Only one possible explanation was left, the engineers said. "That was the first time I heard the words 'defeat device,'" Ayala recalled. "It explained it all." Ayala did not tell anyone at Volkswagen that CARB suspected a defeat device, but he believes that people on his staff probably did.

About the same time, on August 12, Johnson warned his superiors, including Eichler, the head of the emergency task force, and Gottweis, the Red Adair of quality control, that CARB remained dissatisfied. The California regulators "still asked for information," Johnson wrote. "This is not a new request."

With 2016 model year cars still stuck in ports, Eichler assembled a high-level delegation of Volkswagen executives to formulate a plan to get the 2016 model year cars released for sale. The Volkswagen officials still clung to the hope that they could mollify the regulators. They promised to do a second recall of the older vehicles. They assured CARB that they had learned from problems with the older cars and had applied the lessons to the 2016 Volkswagens awaiting approval. The Volkswagen executives provided general information about the software that controlled the emissions equipment, but not the detailed parameters that CARB was seeking. Needless to say, CARB still refused to grant certification. Adding to the pressure, CARB obtained a 2016 model Volkswagen and began making plans to test it, raising the risk that the regulators would make further damaging discoveries.

Volkswagen had run out of excuses. On August 18, Johnson approached Ayala at an industry conference they both attended in Pacific Grove, California, organized by the Institute of Transportation Studies at the University of California, Davis. The gathering took place at a resort known as Asilomar, set amid dunes and pine forests adjacent to a broad sandy beach on the Monterey Peninsula. It is a congenial place for a meeting, offering rustic wood and stone buildings clad in weathered shingles. The attendees included experts from government and academia as well as people from the auto industry. Bosch was a cosponsor. That day Johnson admitted to Ayala that the Volkswagens contained a defeat device. Johnson, who apparently made the confession despite orders from above not to, also informed Christopher Grundler, the director of the Office of Transportation and Air Quality at the EPA, who was also at Asilomar.

Ayala was furious, and he let Johnson know it. He allows he might have used a few obscenities. For well over a year, CARB had been giving Volkswagen the benefit of the doubt, expending countless hours to solve what the company insisted was a technical problem. Now Ayala realized that Volkswagen had knowingly squandered California taxpayer dollars. The company had drained resources that CARB should have been using to help other automakers get their new cars certified. Volkswagen had prolonged the amount of time that polluting vehicles were on California highways. "They wasted our time," Ayala said. "All the testing is very resource intensive. If we're spending all this time on VW, that means we were not taking care of the other things that we could have. We were tying up lab resources. It had a very significant, very real impact on us all."

As word spread inside Volkswagen that the regulators knew about the illegal software, employees began trying to cover their tracks. On

August 31 a meeting was held to discuss how Volkswagen, finally facing the inevitable, would officially admit the existence of defeat devices to regulators. One person present at the meeting was an in-house attorney. He suggested that engineers in attendance should check their documents. Several of those present interpreted the comment as a signal that they should delete anything related to the emissions issue in the United States. One executive, for example, asked assistants to dig out a hard drive containing e-mail correspondence and discard it. In the weeks that followed forty employees at Volkswagen and Audi destroyed thousands of documents. (Many, though not all, were later recovered.)

On September 3, a group of Volkswagen executives, including Eichler, Schmidt, and Johnson, met with CARB officials and formally admitted that the company's vehicles had two calibrations, one for tests and one for normal operation—in other words, a defeat device. In testing mode, the second-generation cars squirted more urea solution, lowering nitrogen oxide emissions. During tests, the cars also increased the amount of exhaust gases recycled into the cylinders, cooling down the combustion temperature and producing fewer nitrogen oxides. But in everyday driving the antipollution measures were scaled back, and nitrogen oxide emissions rose exponentially.

Astonishing as it may seem, Volkswagen was still under the illusion that its belated candor would placate the regulators and get them to release the 2016 cars. Executives still thought they could work out a settlement, fix the cars, and agree to a fine of a few hundred million dollars. But it was far too late for that. Volkswagen had destroyed any chance of an amicable settlement. Ayala's anger was amplified by the realization that the presentation that Johnson and Schmidt had made to him at the conference in Traverse City, Michigan, a month earlier was a fabrication. "For two hours they lied through their teeth," Ayala said, claiming that they admitted as much. "I asked them point blank: 'In Michigan when we met you already knew it was a defeat device?' They said, 'Yeah.'" (Schmidt and Johnson did not reply to requests for their side of the story.)

In defense of its behavior, Volkswagen later maintained that, outside of a small number of people, its executives did not begin to suspect that they were dealing with a possible defeat device until May 2015. Until then, they regarded the emissions problem as purely technical, one of many product issues that car companies routinely face around the world. It is common knowledge in the auto industry, Volkswagen argued, that emissions fluctuate significantly under normal road conditions. Only a small circle of technicians was aware of a defeat device, Volkswagen said, and this group maintained a wall of silence once the company began internal inquiries.

According to Volkswagen's improbable version of events, the problems in the United States were not discussed at length among top executives. Discussion of the diesel issue took only ten minutes of a wide-ranging meeting about product issues on July 21, 2015, the same meeting where executives formed a task force to try to win CARB's approval for the 2016 models. There was no conclusive proof of the existence of a defeat device, Volkswagen maintained, until just before the company confessed to regulators on September 3. According to Volkswagen's account, Winterkorn was definitively informed the next day via a memo in which the issue was the third on a long list of items. The item was just four sentences long. It read in part: "In a meeting on 9/3/2015 with the CARB agency the defeat device (generations 1 and 2) was admitted. The agency had in addition inquired about similar functions in the Audi V6 TDI." (Winterkorn declined to comment or answer any other questions from the author.)

The proverbial oil well fire that Gottweis had warned Winterkorn about more than a year earlier was now out of control, and not even Red Adair could put it out.

CHAPTER 18

Empire

Usually the Ballsporthalle in the working-class Frankfurt suburb of Hoechst was the home of the Frankfurt Skyliners, a mediocre pro basketball team. On September 14, 2015, it was repurposed into a showcase for all that Volkswagen had become since Herr Porsche unveiled the original people's car to the German people in 1938.

Outside the entrance, a pair of tractor trailers stood sentry, products of Volkswagen's MAN and Scania truck divisions and the first reminder that the company had long since become a global producer of practically everything with wheels and a motor. A slow-moving caravan of chauffeur-driven Porsche SUVs deposited representatives of the motoring press, politicians, industry analysts, and other important people at the doorway. Inside, guests jostled at the open bars for wine, beer, hot dogs, and miniature tacos until they were shooed toward the courtside bleachers, which quickly overflowed. Latecomers had to sit in the aisles.

It was the eve of the Frankfurt International Motor Show, when traditionally Volkswagen's top managers take the stage to show off their shiniest new models to the media and selected guests. Auto executives today are expected to be not only experts in production, supply chains, marketing, and design but also impresarios, able to muster the required stage presence when it is time to roll out a new model before a car show audience. So it was on this night, as one by one the managers of Volkswagen's car divisions, all eight of them, took the floor. They spoke German and read their texts from a video screen, speaking quickly because there was much to show off and only an hour in which to show it. Wolfgang Dürheimer, the chief executive of Bentley, was up first. Ferdinand Porsche would certainly have been surprised, and perhaps amused, to discover that Volkswagen was now the owner of this symbol of the British upper class. But he would have understood the logic. Bentley was proof that Volkswagen could serve fussy rich people as well as common folks. Dürheimer unveiled a gold-painted Bentayga, which, he boasted, was "the most exclusive, the most powerful, and the fastest SUV in the world." He did not explain who would be crazy enough to drive a $230,000 car in the mud, but never mind. The point was that the most over-the-top SUV in the world came from Volkswagen.

Next came Stephan Winkelmann, the chief executive of Lamborghini, the Italian maker of super sports cars, another acquisition that might have startled Porsche. Fit-looking, his longish brown hair impeccably coiffed, Winkelmann greeted the audience with a perfectly accented "buona sera," before switching to German. He was clearly more comfortable in the spotlight than some of his more Prussian coexecutives, and his contribution to show and tell was the Huracán LP 610-4 Spyder, a soft-top monster with a 610-horsepower, ten-cylinder motor and a carbon fiber body. Yet lest anyone think that the Huracán was merely an expensive toy, Winkelmann pointed out that it emitted 14 percent less carbon dioxide than its predecessor. In line with Volkswagen's commitment to the environment, even Lamborghini was doing its bit for the planet.

The sheer breadth of the Volkswagen product palette would have been astonishing to Herr Porsche. It was not just a car company; it was an empire. There was a SEAT from Spain, a Skoda from the Czech Republic, a Ducati motorcycle, and—the final affirmation that Volkswagen was no longer just a provider of practical transportation for the middle class—a million-dollar-plus Bugatti. There was, to be sure, even a Volkswagen, a Tiguan SUV, introduced by Heinz-Jacob Neusser, head of development for the Volkswagen brand. Though each badge had its own style, they had much in common under the skin. To save money and undercut competitors, Volkswagen deployed components like engines, transmissions, and chassis among the different brands. Even the Lamborghini shared parts with its plainer cousins, though only a mechanic would be able to tell.

Ferdinand Porsche wouldn't have been completely surprised by the latest Porsche 911, introduced by Wolfgang Hatz, chief of development for the sports car maker. The 911—a variant known as the Carrera S—still betrayed a genetic similarity to the original 356 designed by Porsche's son, Ferry. Then and now, Porsches were distinguished by the line of the hood and the way it sloped below two raised headlights. The design gave Porsches a look of alertness. But Ferdinand Sr. might have been puzzled to learn that Porsche was now a division of Volkswagen. The two companies had long collaborated on design and production, but always at a certain remove. And he would have been further baffled to hear Matthias Müller, the Porsche chief executive, talking about a new sports car prototype that would run completely on battery power.

Ferdinand Porsche's grandson Ferdinand Piëch, the man largely responsible for assembling all the brands on display, was not in the room that night. He had avoided the spotlight since losing the power struggle with Winterkorn a few months earlier. The Porsche family was represented instead by Wolfgang Porsche, Piëch's avuncular cousin and longtime antagonist. The evening belonged instead to Winterkorn, a man very much indoctrinated in the Porsche way of thinking even if he had recently survived a nasty falling out with the old man's most prominent grandson. That is to say, Winterkorn occupied a world where engineering excellence was the highest art, and worth many sacrifices to achieve.

Winterkorn spoke last, about an hour after the presentations had begun. An announcer introduced him as Professor Doctor Martin Winterkorn. The implication was that he was a man not only powerful but learned.

Winterkorn's appearance as the last presenter reflected not only his rank but also his intense personal involvement in Volkswagen products. In Germany, it was said that Winterkorn knew every screw in every vehicle produced by the Volkswagen Group, all three hundred models. It was a company where top management could have a long discussion about the proper angle of a windshield. An oft-published photograph of Winterkorn showed him crawling underneath a car, still wearing a suit, to inspect the undercarriage.

To veterans of Volkswagen Group Night, as the preshow extravaganzas were somewhat blandly known, the one in Frankfurt in 2015 was a touch less bombastic than usual. There was, for example, no surprise appearance by Pink, the pop singer, as there had been at a Group Night a few years earlier. Still, there was loud music, heavy on drums and synthesizer, and a laser light show. At one point, the laser projector conjured the image of a street and buildings, which created the illusion that a Volkswagen Golf sedan—a real one driving across the stage—was traveling through an urban landscape.

When Winterkorn finally took the floor, the laser projector bathed him in blue light, Volkswagen's signature color. As soon as he began to speak, it was clear that something was different. Winterkorn talked about how Volkswagen was shifting its energies to plug-in hybrids and battery-powered cars, instead of the diesel engines that had formed such a core part of the company's marketing strategy in Europe and the United States. A fossil fuels guy of long standing, Winterkorn mentioned the word "diesel" just once. Instead, he promised, Volkswagen would bring twenty new plug-in hybrids and electric cars to market by the end of the decade. Winterkorn alluded to the Silicon Valley companies that were beginning to take an interest in the car business—a reference to Google, which was devoting some of its huge financial resources to develop cars that could drive themselves, and to

Apple, said to have assembled a large team to design an electric car. Volkswagen, too, would develop self-driving cars, he said, and they would be "not just for the upper class, but for everyone." What's more, all Volkswagen Group vehicles would soon be smartphones on wheels. (The statement made at least a few well-informed listeners chuckle. A few days earlier, Müller, the Porsche chief executive, had told an auto magazine that his vehicles would not be smartphones on wheels.)

It was a sudden conversion for a company that had long been disdainful of hybrids and battery-powered cars. Volkswagen was the only major carmaker that refused to join the California Plug-In Electric Vehicle Collaborative, a joint effort by manufacturers, utilities, non-profit organizations, and governments to promote use of battery-powered vehicles. But, Winterkorn said, horsepower was no longer the decisive competitive factor in the auto business. "Ever faster, higher, farther, is no longer enough," he said. That was, to put it mildly, an unexpected change of attitude. During the two decades that Winterkorn and Ferdinand Piëch had managed Volkswagen together, pushing the boundaries of automobile engineering was the whole point. Volkswagen would become the biggest company. It would build the fastest production car (the Bugatti Veyron). It would build the most beautiful factory (the glass-walled plant in Dresden where Volkswagen made the luxury Phaeton). It would build the first four-door production car that could travel 100 kilometers on less than three liters of fuel (the short-lived aluminum-bodied Audi A2). And Volkswagen would build the cleanest diesel engine.

Most of the people in the audience had no inkling, of course, of the scandal that was about to break over Volkswagen like a tidal wave. But those familiar with Winterkorn might have noticed that something about his demeanor was off. He seemed less confident, less imperial than usual, almost emotional. Winterkorn's restrained affect seemed odd considering that, the previous July, Volkswagen had achieved its long-cherished goal of producing more cars worldwide than Toyota.

By surpassing Toyota, Volkswagen demonstrated how much it had conquered the world. In China, Volkswagen sold more cars than any rival, with a 21 percent share of the passenger car market, ahead of number two–ranked General Motors. Volkswagen was also strong in Latin America, with a local manufacturing network that included three large car factories in Brazil. Volkswagen was a leading player in all of the world's largest car markets except for the United States, where the VW brand was struggling. Sales were down 2.5 percent so far that year in an otherwise booming U.S. market. Winterkorn did not mention any of that.

But he made it clear that Volkswagen was not ready to give up on the United States. The company planned to produce a new crossover SUV at its factory in Chattanooga, addressing the American fondness for four-wheel drive vehicles. "Volkswagen is staying in the driver's seat," Winterkorn said.

FOUR DAYS LATER, on September 18, 2015, the Environmental Protection Agency held a press conference and issued a so-called notice of violation accusing Volkswagen of using software "that circumvents emissions testing for certain air pollutants." An EPA news release said that regulators had found nitrogen oxide emissions as high as forty times the limit in some Volkswagen vehicles. The EPA informed Volkswagen of the announcement only half an hour beforehand.

The press release left no doubt about the severity of the violation. Nearly 500,000 cars in the United States were affected, exposing Volkswagen to a maximum fine of $18 billion. Even if Volkswagen escaped the maximum penalty, the cost was sure to make the $100 million fine paid recently by Hyundai-Kia look like a traffic ticket. The health consequences of Volkswagen's misconduct were serious, the EPA said. Nitrogen oxide pollution "has been linked with a range of serious health effects, including increased asthma attacks and other respiratory illnesses that can be serious enough to send people to the hospital."

The pollution from the cars contributes to the formation of smog and produces soot that has "also been associated with premature death due to respiratory-related or cardiovascular-related effects," the agency said. "Children, the elderly, and people with pre-existing respiratory disease are particularly at risk for health effects of these pollutants."

Elizabeth Humstone, a sixty-seven-year-old urban planner, learned about the accusations against Volkswagen from a news alert on her smartphone. Humstone lived most of the time in Charlotte, Vermont, a quiet village on the eastern shore of Lake Champlain. She was chairwoman of the board of directors of the Vermont Natural Resources Council, an environmental group. She was also the owner of a 2011 Volkswagen Jetta SportWagen TDI diesel—one of the customers whose loyalty had helped build the Wolfsburg empire.

Humstone had spent essentially her entire career involved in environmental and planning issues. She boasted that she came from a family that was recycling trash in the 1950s, and she had a master's degree in city planning from Harvard University. Later she was executive director of the Vermont Forum on Sprawl, which tried to keep the state's open spaces from being covered by McMansions.

When Humstone began shopping for a new car in 2010, she thought first of a Prius, the Toyota hybrid that pioneered the idea of a car that was less harmful to the planet. The Prius was the default choice for any people who considered themselves environmentalists. "All my work was related to being environmentally responsible," Humstone recalled later. "That was really important to me. If I was going to purchase a car, I wanted one that was responsible."

Humstone test-drove a Prius. But after consulting her circle of friends, she also visited a dealer in South Burlington, Vermont, and drove a Volkswagen diesel. The Prius saves fuel and reduces pollution by combining electric and gasoline motors. But the diesel Jetta also had good mileage numbers, 42 mpg on the highway compared with 48 for the Prius, according to EPA ratings. Many Volkswagen owners reported much better mileage in everyday driving.

Humstone spent a lot of time driving between her partner's home in Concord, Massachusetts, and her own in Vermont, 220 miles away. She also spent a lot of time in Maine. She needed a car with good highway economy. She had seen the "clean diesel" ads. And she liked the Jetta's performance. "I liked the gas mileage from the Prius, but it wasn't as responsive, it wasn't as peppy," Humstone said. "Knowing I had these long-distance drives, I decided to go with a car I thought would be more fun." She paid about $29,000 for the Jetta, not counting a federal tax deduction she received for buying a fuel-efficient car.

ON SEPTEMBER 18, the day that a *New York Times* headline popped up on Humstone's phone, she was in Concord, Massachusetts, visiting her partner. It was a Friday. Humstone realized that clean diesel was a lie, and she was livid. She was now stuck with a car that instantly became synonymous with environmental fraud. Soon thereafter, she went on Facebook. "Are other VW diesel car owners out there concerned about driving around in a vehicle polluting the air?" Humstone wrote. "What are you doing about it? Are there citizen action groups being formed? I feel I was sold a car on very false pretenses thinking that I was taking an environmentally responsible step for better gas mileage while meeting EPA standards. Now I feel guilty every time I drive."

Millions of other people around the world shared Humstone's outrage. It soon became clear that Volkswagen was involved in one of the greatest corporate scandals ever. Between 2009 and 2015, the company had sold 11 million cars with software designed to deceive the emissions police. That included nearly 600,000 cars in the United States, 2.8 million in Germany, and 1.2 million in Britain. It was the automotive equivalent of Enron, the once-admired energy company that collapsed in 2001 after much of its profit turned out to be fictitious; or the Libor scandal, in which Deutsche Bank, Credit Suisse, and other banks colluded to rig benchmark interest rates in order to increase their trading profits. But while those cases involved financial crimes—theft, essentially—and were driven by obvious greed, the Volkswagen case was in a different category. The financial gain was modest, especially in light of the eventual cost. By 2012, the price of SCR systems had fallen enough that they were about $50 more expensive per car than a lean NOx trap, according to the ICCT. (To combine the two systems, as BMW did, would have added about $500 to the cost of a vehicle.) In contrast to the bank scandals, which sometimes generated bonuses worth tens of millions in dollars or euros to their perpetrators, the extra profit generated by the emissions cheat did not flow to any of the individuals directly responsible. They were motivated by something other than money. The mystery was why the people responsible would take such an enormous risk for such a modest gain.

That same day, Dan Carder was working in the cinder block building on the West Virginia University campus used for testing engine emissions. The EPA press release a few hours earlier had mentioned the work done by WVU to expose Volkswagen's cheating, but Carder didn't know that yet. He was wearing jeans and a T-shirt and was covered in grease. His mobile phone was sitting on a windowsill, recharging. When Carder checked the phone, he noticed that he had a slew of missed calls from area codes all over America. Then the phone rang. It was a reporter from one of the television networks. "Do you know why we're calling you?" the reporter asked.

Carder's first thought was that one of the university's mobile emissions testing labs, which roamed the country on trailers, had had an accident. "I'm going through my head, where do we have people in the United States? Where are the crews at? Did something just blow up? What did we do wrong?" He was not aware of the investigatory work done by CARB and still believed that the sky-high emissions his team had detected in Volkswagens two years earlier were the result of mechanical defects or a design flaw. Carder was floored to learn that Volkswagen had deliberately tried to deceive regulators.

At least since 1998, when truck engine makers paid one billion dollars to settle an emissions scandal, vehicle manufacturers had been on notice that getting caught trying to fool the government could be expensive. In the meantime, the technology to check auto emissions on the road had also advanced, increasing the chances of discovery. Everyone in the industry knew that—except, apparently, Volkswagen. "I was shocked," Carder said. "After all the stuff that happened in the nineties, how could anyone do this?"

CHAPTER 19

Aftermath

THE INTERNATIONAL MOTOR SHOW in Frankfurt had been underway for three days when news of the scandal broke. The show, held every two years at a vast exposition center on the edge of the Frankfurt city center, bills itself as the world's largest. The event is traditionally an occasion for German automakers to revel in their country's disproportionate share of the high end of the car market. BMW, Mercedes-Benz, and Volkswagen rent entire exhibition halls. Their displays make such rivals as Renault, Opel, or Fiat Chrysler look like boutique automakers. Among the 930,000 car enthusiasts who flooded the grounds was Angela Merkel, the German chancellor, who stopped by the Volkswagen Group stand and posed for photos with the Porsche Mission E prototype. In the auto world, news of the scandal was an acute embarrassment, the equivalent of throwing a lavish party to impress all your friends and having the cops show up.

The EPA issued its news release about Volkswagen just before the motor show swelled with weekend crowds. The headlines about Volkswagen's wrongdoing soured the fun and made everyone look bad. "There was a certain mood hanging over the show," said Matthias Wissmann, the president of the German Association of the Automotive Industry, which organized the motor show. "Especially among the professionals, the worry was: what happens now?"

The essence of the "made in Germany" brand is that, whatever people might think of the Germans, they're damn good engineers. In Volkswagen's case, it now looked as though some of the vaunted engineering was just bluff.

The ferocity of the public reaction appeared to catch Volkswagen by surprise. There was no plan for reacting to the media firestorm. Several members of the supervisory board, including Stephan Weil, the prime minister of Lower Saxony, later complained that Volkswagen management never informed them about the emissions problem and that they learned about it from the news media. It took two days for Winterkorn to issue a statement. He expressed contrition and promised to cooperate with investigators. "I personally am deeply sorry that we have broken the trust of our customers and the public," Winterkorn said.

The financial consequences were immediate. The following Monday, the first full stock trading day after the announcement, Volkswagen shares fell 20 percent. Many of the big sellers were mutual funds that focused on companies with good environmental track records. Volkswagen was dropped from the Dow Jones STOXX ESG Leaders 50 index, a listing of companies seen as paragons of sustainability and good governance. Funds that tracked the index by holding the same shares then automatically sold their stock in Volkswagen. By the end of the month Volkswagen shares were down 45 percent.

Other authorities followed the EPA in pursuing legal action against Volkswagen. Germany's financial markets regulator, known by its German acronym BaFin, began an investigation of whether Volkswagen had broken securities laws by failing to notify shareholders sooner of the emissions issue. Prosecutors in Braunschweig, near Wolfsburg, opened a criminal investigation, as did the Department of Justice in the United States, as well as authorities in countries like France, Britain, and South Korea. The European Union's antifraud office began investigating whether Volkswagen had misused hundreds of millions of euros in low-interest loans provided by a publicly funded development bank. Lawyers in the United States announced plans to file class action suits on behalf of Volkswagen owners.

The allegations by the EPA and CARB turned out to understate the scale of the fraud. Volkswagen admitted on September 22 that all eleven million cars with the EA 189 motor worldwide had the illegal software. The case illustrated the downside of Piëch's strategy of sharing parts among as many vehicles as possible. One problem infected every car with that component, creating a global disaster. To cover the anticipated cost, the Volkswagen supervisory board set aside €6.5 billion, or $7.3 billion. It was obvious that that sum would not be nearly enough.

Winterkorn tried for several days to hang on to his job. A person who saw him in Wolfsburg during those days said he looked shattered by the turn of events. But on September 23, 2015, five days after the EPA news release, Winterkorn surrendered to the inevitable and resigned. "As CEO I accept responsibility for the irregularities that have been found in diesel engines," Winterkorn said. His statement was no mea culpa, though. He maintained that he was as surprised by the revelations as anybody. "I am shocked by the events of the past few days. Above all, I am stunned that misconduct on such a scale was possible in the Volkswagen Group," Winterkorn said. "I am not aware of any wrong doing on my part."

Winterkorn had had less than five months to manage Volkswagen without Piëch's looking over his shoulder. Now the two people who had dominated the company since the 1990s were gone. But Winterkorn's departure did not signal a broad management shake-up. Volkswagen suspended some lower-ranking executives, with pay, who had held key positions in engine development. They included Heinz-Jakob Neusser, head of development for the Volkswagen brand, who had presented a new version of the Tiguan compact SUV at Group Night; and Wolfgang Hatz, the head of research and development at Porsche, who had unveiled a new 911 sports car. Hatz had been head of engines and transmissions development for the Volkswagen Group in 2007; he was the same person captured on video several years earlier complaining about California emissions regulations. Christian Klingler, head of sales and marketing, was the only member of the management board besides Winterkorn to leave the company in the days following the EPA accusations. His resignation was "a result of differences with regard to business strategy," Volkswagen said, and was "not related to recent events." The rest of the management board stayed in Wolfsburg.

Two days after Winterkorn's resignation, the Volkswagen supervisory board named Matthias Müller, the chief executive of Porsche, to replace Winterkorn. Ironically, Müller had been Piëch's rumored favorite to replace Winterkorn the previous April. It appeared that Piëch had inadvertently gotten his wish.

Müller, sixty-two years old, was a less forbidding figure than his predecessor. He was neither as loud as Winterkorn nor as coldly intimidating as Piëch. Slender and fit-looking, with longish gray sideburns, Müller projected a demeanor of cool confidence. As befit the chief executive of Porsche, he was a racing enthusiast, a car guy skeptical of the new infatuation with digitally connected vehicles. Müller had a reputation as a ladies man. The gossip magazine *Bunte* linked him to Barbara Rittner, a former professional tennis player turned coach. They had met at tennis events sponsored by Porsche. Müller, himself an enthusiastic tennis player, was still married to the mother of his two grown children. But in Germany—or at Volkswagen, at any rate—that hardly disqualified him as a chief executive.

Otherwise Müller came across as more compassionate and socially conscious than Winterkorn or Piëch. They rarely commented on matters outside the world of cars. Müller, by contrast, publicly endorsed Chancellor Merkel's open-door policy toward refugees from Syria who were then flooding into Europe. In an interview with the *Süddeutsche Zeitung* before he became chief executive, Müller likened the refugees' plight to that of his own family, which had fled East Germany when he was three. The Syrians had it much worse, Müller said, because they had to cope with a completely unfamiliar language and culture. "We have to help," he said. That was not a statement one would have expected to hear from Winterkorn or Piëch.

One of Müller's first official acts was to acknowledge that Volkswagen had an outmoded system of management. The company was not a "one-man show," he told a gathering of Volkswagen managers, in an apparent reference to the dominant personalities who had preceded him. "A company of this size, this internationality, and this complexity cannot be managed with structures and principles from yesterday," Müller said. "A new corporate structure is overdue."

Müller promised to decentralize the organization and give managers of Volkswagen brands more autonomy. "I am definitely not planning to intervene in product decisions," Müller said, in an obvious reference to Winterkorn's penchant for micromanaging every aspect of car design. "Whether a front windshield stands one degree steeper or not—I'm not going to concern myself with that." In a signal that he would be less imperial than Piëch or Winterkorn, Müller declined to take up residence in their former offices in the executive high-rise in Wolfsburg—the building capped with a giant VW hood ornament. Müller kept his offices in a low-rise building nearby.

But Müller was hardly a new broom. He was an insider who had spent his entire career at Volkswagen, joining Audi out of secondary school as an apprentice toolmaker. He left the company to earn a degree in computer science at a technical college in Munich, but returned afterward and made a name for himself in the 1990s as product manager for the A3, Audi's entry-level compact wagon. In 2007, Winterkorn named Müller head of product management for Volkswagen, the same period when the company was struggling to solve the diesel emissions problem and resorting to illegal means. While promising to be less autocratic than his predecessor, Müller spoke admiringly of Winterkorn. "I have huge respect for his achievements," Müller said soon after becoming chief executive.

Any doubt that insiders were still in control in Wolfsburg was erased by the appointment of a permanent successor to Piëch as chairman of the supervisory board. Shareholders elected Hans Dieter Pötsch, who had been chief financial officer under Winterkorn. Pötsch's nomination for the job was announced several weeks before the emissions scandal came to light but took effect afterward. An Austrian, Pötsch was close to the Porsche family and regarded as a guardian of the interests of the majority shareholders. The old guard remained very much in control.

To the extent that Volkswagen had a communications strategy, it seemed to consist of apologizing effusively for the wrongdoing, while insisting that no one at the top had known about it. A few days after the EPA disclosures, Michael Horn, chief executive of Volkswagen of America, spoke at a previously scheduled event in New York to launch a redesigned Passat. "We screwed up," Horn told the crowd. Several weeks later, he appeared before a congressional subcommittee, apologizing again but implying that the cheating was the work of rogue underlings. "On behalf of our company, and my colleagues in Germany, I would like to offer a sincere apology for Volkswagen's use of a software program that served to defeat the regular emissions testing regime," Horn told the U.S. representatives on October 8, 2015. Under oath, he allowed that he had been informed in the spring of 2014 about "possible emissions non-compliance." But he said he had been assured that Volkswagen engineers were working to fix the problem in cooperation with regulators. "I did not think that something like this was possible at the Volkswagen Group," Horn stated.

After becoming chief executive, Müller adhered to the same line. "Based on what I know today only a few employees were involved," Müller told the *Frankfurter Allgemeine* newspaper. He also defended Winterkorn, with whom he had worked closely. Despite Winterkorn's reputation for being immersed in the workings of Volkswagen products, Müller said, that attention to detail did not extend to computers. Müller doubted that Winterkorn knew about the cheating. "Do you really think that a chief executive had time for the inner functioning of engine software?" he asked the newspaper.

Volkswagen's supervisory board hired the U.S. law firm Jones Day, which has a large presence in Germany, to conduct an internal investigation, which it promised would be exhaustive and independent. Wolfgang Porsche, Ferdinand Piëch's cousin, was put in charge of a supervisory board committee overseeing Jones Day's work.

As weeks passed, Volkswagen's problems with regulators in the United States continued to deepen. On November 2, 2015, the EPA issued another notice of violation, this time for cars with the three-liter diesel motors. The vehicles affected included SUVs like the Volkswagen Touareg, Audi Q5, and Porsche Cayenne, as well as several larger Audi passenger cars. According to the EPA, the three-liter engines also had defeat devices.

About 80,000 cars in the United States were affected by the second notice of violation. That was far fewer than the 500,000 vehicles, all Volkswagen brand cars except for the Audi A3, covered by the first EPA citation. But the second group consisted mainly of luxury cars, typically selling for $50,000 or more. In the case of the Porsche Cayenne SUVs, the sticker price could easily top $100,000 with options. The danger to Volkswagen from the accusations against Audi and Porsche was greater than the number of cars would suggest. The United States had been Porsche's most important foreign market since the 1950s. Audi had been doing well in the United States in recent years, remodeling dealerships and gaining ground on BMW and Mercedes-Benz. Volkswagen did not disclose profits for its business in the United States, but it is likely that the pattern was the same as in the rest of the world. Audi and Porsche were the main sources of profit, and the margin on Volkswagen brand cars was slim or nonexistent. Volkswagen could not afford for Audi and Porsche to be tainted by scandal as well.

The emissions revelations quickly began to hurt sales worldwide. The decline was most immediate in the United States, in part because Volkswagen halted sales of diesels, which had accounted for about one-fifth of the volume. But the damage to the company's reputation obviously didn't help. Sales were stable in October 2015, the first full month of sales after the EPA citation. But in November sales of Volkswagen brand cars plunged 25 percent in the United States, even as the overall auto market boomed. In December, sales fell another 9 percent, even though a redesign of the Chattanooga-made Passat was arriving at dealer lots. Normally, a new version of a popular model would give sales a significant boost. Volkswagen attempted to preserve customer goodwill by offering owners in the United States a prepaid Visa card they could use to make $500 in purchases, another card good for $500 at Volkswagen dealers, plus free roadside assistance. But many owners regarded the gesture as laughably inadequate.

"So basically VW is offering a $500 gift for the fraud it perpetrated on me and its other customers, the dealers and the US government," said Elizabeth Humstone, the Vermont environmental activist. "I have been driving my 2011 Jetta Sportwagon TDI for just over five years. During that time, I have driven 88,000 miles spewing nitrous oxide in excessive amounts, but thinking I was doing the right thing for the environment. VW has a long way to go to make me feel any better."

The damage to sales in Europe was less dramatic, but potentially more threatening to Volkswagen's bottom line. Europe is Volkswagen's core market, accounting for about 40 percent of sales volume. Even the loss of a few points of market share can be serious. During the last few months of 2015, Volkswagen's overall sales in Europe grew, but at a much slower rate than those of its rivals. In December, Volkswagen sold 5 percent more cars than a year earlier. Renault, Ford of Europe, and PSA Group, the maker of Peugeot and Citroën cars, all recorded gains of more than 20 percent for the month, as the European car market rebounded from a severe downturn. So even though overall sales in Europe rose, Volkswagen's market share for all its passenger car brands fell in December 2015 to 22 percent, from 25 percent a year earlier. Adding to the problems in Europe, countries like Switzerland had banned sales of Volkswagen diesels shortly after the cheating came to light.

Even before the emissions scandal, Volkswagen was in a weakened state. An economic slowdown had hurt sales in China, which had become the single largest market for Volkswagen. Sales in Brazil, another important market, were plunging because of that country's economic crisis. In Russia, at one time a promising growth market, sales were suffering because of sanctions imposed by the United States and Europe in connection with the conflict in Ukraine. In addition, Russian purchasing power was undermined by slumping prices for oil and other commodities, the basis of the economy. Volkswagen's ability to absorb these setbacks was hurt by its persistent efficiency gap in relation to Toyota. In October, Volkswagen reported its first quarterly loss since it had begun disclosing three-month figures fifteen years earlier. The loss was a direct result of the emissions scandal: accounting rules required the company to subtract from profit the money set aside to cover costs stemming from the emissions fraud. The nine-month sales figures also showed that Volkswagen had slipped back into second place in terms of the number of cars sold worldwide. Müller, though, said that size was no longer an end in itself—repudiating Piëch and Winterkorn's long-held dream of world dominance. The quest for bigness had skewed Volkswagen's priorities, Müller argued. "A lot of things were subordinated to the desire to be 'faster, higher, larger,'" he said.

Wavering sales and the prospect of billions of dollars in fines, court judgments, and recalls provoked a financial chain reaction. In November 2015, Fitch Ratings was the first of the major ratings agencies to downgrade Volkswagen's creditworthiness. "We believe that the emergence of a fraud of this magnitude, going either unnoticed or uncorrected by top management for so long is not consistent with" a top credit rating, Fitch said. Shortly afterward Standard & Poor's and Moody's also cut Volkswagen's ratings. Volkswagen's declining creditworthiness presented yet another threat to profits, because it meant that investors might demand higher interest rates on Volkswagen debt and raise the company's financing costs. Volkswagen would not be able to pass the higher rates on to car buyers without hurting sales.

Like most big automakers, Volkswagen has a large financial services arm. The unit sells bonds and other forms of debt to investors, raising money that the company lends to customers so they can buy or lease cars. In good times, financial services are a major source of earnings for Volkswagen, generating more profit than the company's MAN and Scania truck divisions combined.

Legally, Volkswagen Financial Services is a bank, one of Europe's largest, with about €114 billion, or $125 billion, in loans and other assets at the end of 2015. That figure does not include Volkswagen's separate finance unit in the United States, VW Credit Inc., which had a $28 billion loan portfolio, mostly in the form of credit to people who bought or leased vehicles. As a bank, Volkswagen Financial Services is considered important enough to be overseen directly by the European Central Bank, rather than solely by local regulators. The unit is among 129 banks, out of about 5,500 in the eurozone, deemed worthy of special central bank scrutiny because they could damage the whole financial system if something went wrong. The only other auto company bank to earn that distinction is RCI Banque, Renault's finance unit.

Volkswagen was also exposed to the financial fortunes of many of its dealers, especially in Germany. It had lent them €10 billion, or about $11 billion, for purposes such as remodeling dealerships. If sales slumped, some dealers might get into trouble and not be able to repay their loans to Volkswagen, creating a vicious circle. "The performance of the dealer loan book is to a large extent related to the well-being of the related car manufacturer and its brands," Moody's said.

Volkswagen Financial Services' biggest concern was probably the hit it could take from leased vehicles. Usually Volkswagen is obligated to take back leased cars when the contracts expired. If the value of the vehicles dropped more than planned, Volkswagen would have to absorb the difference. That risk prompted the financial services unit to subtract €450 million from profit in 2015, or about $495 million, in order to set the money aside as insurance against possible losses from leased cars. In fact, resale values of Volkswagens were already plummeting in the United States, which accounted for about a fifth of the leasing contracts. Kelley Blue Book, which tracks used car values, estimated that prices for used Volkswagens had by the beginning of 2016 fallen 16 percent, from their pre-scandal value.

Some owners reported much steeper losses. Mark Winnett, a drug development team leader for a pharmaceutical company, was forced to sell his 2014 Jetta in December 2015 when he moved from Phoenixville, Pennsylvania, to a new posting in Shanghai. Winnett, who was fifty-eight, paid about $28,000 for the car in mid-2014. A Volkswagen dealer took the car off his hands for a little more than half that amount. And the dealer acted as if he were doing him a favor, Winnett said.

"I felt betrayed," Winnett admitted. He was among the people who believed Volkswagen's claims to be environmentally friendly. He also owned a battery-powered Nissan Leaf. "We drove the Jetta for almost a year and a half thinking not only were we driving a great car but also reducing emissions," he added.

It was fortunate for Volkswagen that the scandal came to light when European interest rates were at all-time lows. The European Central Bank was effectively printing money in an attempt to stimulate the moribund eurozone economy. The monetary policy measures pushed rates for highly rated corporate debt close to zero, and kept Volkswagen's borrowing costs low even after the downgrades. Still, there were indications that investors would demand higher interest rates to buy Volkswagen debt than they might have otherwise. In November, Volkswagen sold asset-backed securities tied to leased vehicles in Germany. The securities are bundles of debt whose purchasers are entitled to proceeds from the payments. Volkswagen had to pay investors about half a percentage point more than it had the previous May for similar debt.

Volkswagen had plenty of cash and was not in danger of going bankrupt, at least not right away. But the financial burden forced the company to cut back on research and development just when the car industry was on the cusp of major technological change. In November, the company said it would reduce spending on new models and other projects by about €1 billion, to €12 billion, or $12.8 billion. It was the first time Volkswagen scaled back investment since 2009. Among the projects to be cut was an update of the Phaeton. Production at the Transparent Factory in Dresden was suspended in early 2016, and most of the workers were transferred to other sites. The plan, ostensibly, was to retool the factory to produce a new luxury battery-powered vehicle that had not yet been developed and, in any case, would not be ready until 2020. The last Phaeton rolled off the assembly line on March 18, 2016. Dozens of workers in their white uniforms crowded around the black sedan for a last group photo.

While probably no one but Ferdinand Piëch would rue the unpopular luxury Volkswagen, the cuts came as the auto industry faced new competition from Silicon Valley. Apple was rumored to be working on its own car project. Google was investing massively in self-driving cars. At auto shows, the talk was of a move to battery power. Though the number of electric cars on the roads was still tiny, executives worried that a shift away from internal combustion engines to batteries would make them vulnerable to new competitors, perhaps from China. Volkswagen would have less money to respond to all these changes.

By December 2015, the internal investigation overseen by Jones Day had expanded into a massive undertaking. As Volkswagen liked to point out, it involved 450 external and internal experts who had secured 100 terabytes of data, the equivalent of 50 million books. The emphasis would be on thoroughness, not speed, the company said repeatedly. The implication was that it would take a long, long time to find the culprits.

In the meantime, Volkswagen disclosed only limited information about what Jones Day was uncovering. In December, Müller and Pötsch held a news conference in Wolfsburg to present what they described as preliminary findings of the inquiry. The event took place in the AutoUni, a company training center a short walk from the main factory. The aggressively modernistic AutoUni was another architectural manifestation of Volkswagen wealth and power. The building consisted of two modern connected structures made of concrete, glass, and steel, and shaped like a pair of huge parallelograms lying on their sides. Müller and Pötsch sat side by side at a large white desk with "Volkswagen" printed in large gray letters on the front. Speaking to several hundred journalists, they conceded that the emissions fraud was the result of "weaknesses in some processes" and "a mindset in some areas of the company that tolerated breaches of rules." They promised to improve the procedures for checking and approving software, which, they said, "were not suited to preventing use of the software in question."

But the chief executive and the supervisory board chairman stuck to the same basic story Volkswagen had promulgated since the emissions cheating first came to light. They blamed it on "the misconduct and shortcomings of individual employees." Nine Volkswagen employees had been suspended in connection with the wrongdoing. Volkswagen did not disclose any of their names, though most of them had already leaked out. They included Frank Tuch, the head of quality control; and Ulrich Hackenberg, a member of the Audi management board responsible for technical development, who had already resigned his job. Müller and Pötsch did not address the fifteen-month cover-up that had occurred following the West Virginia report. They certainly did not leave the impression that the emissions scandal could be the result of a broad conspiracy that reached to the top of the company.

CHAPTER 20

Justice

THE VAST MAJORITY of Volkswagens with tainted software were in Europe, but the company's biggest legal problems, by far, were in the United States. The difference was enforcement. Even though European law also prohibited defeat devices, the rules were vague and there were practically no penalties. The worst that could happen, or so it seemed, was that Volkswagen would have to fix the cars, a much easier task in Europe because its regulations allowed nitrogen oxide levels to be more than twice as high as those permitted in the United States. In fact, in November 2015, Volkswagen announced a plan it said would make the 8.5 million cars in Europe compliant with EU clean air rules at a relatively modest cost. The fix consisted of nothing more than a software update and, in some models, installation of a so-called flow straightener—a tubular plastic part, slightly larger in diameter than the cardboard tube inside a roll of paper towels. The part was supposed to improve air flow into the engine and reduce emissions. The fix was so simple it raised the question of why the company cheated in the first place. Volkswagen's only explanation was that its understanding of the technology had advanced since the cars were built. Environmental groups were skeptical, but German regulators certified the fix. According to EU rules, German approval automatically made the modification legal in all the member countries.

Volkswagen was not going to get off so easily in the United States. With its fraudulent clean diesel push, followed by an obfuscation campaign, Volkswagen had exposed itself to the full wrath of the American justice system. The company could hardly have picked a worse time to get caught. The EPA's formal notice of violation against Volkswagen happened to come only nine days after the U.S. Department of Justice had declared a new resolve to hold top executives accountable for corporate wrongdoing. Sally Quillian Yates, a deputy U.S. attorney general, issued a memo on September 9, 2015, instructing the department's lawyers not to agree to settlements with corporations accused of wrongdoing unless they also included punishment for the people responsible. The memo, which attracted wide attention, made it clear that the government should not be content with nabbing a few middle managers. Investigators should target "high-level executives, who may be insulated from the day-to-day activity in which the misconduct occurs."

The Yates memo was inspired by the global financial crisis, in which widespread wrongdoing by banks caused enormous economic pain but led to only a handful of criminal convictions. Shareholders often wound up bearing the financial burden of fines, while managers walked away richer. The Yates memo, addressed to top Justice Department lawyers as well as to the director of the FBI, was a call to rectify that injustice. "One of the most effective ways to combat corporate misconduct is by seeking accountability from the individuals who perpetrated the wrongdoing," Yates wrote. That was a polite way of saying that the best way to get the attention of executives was to put a few behind bars.

Volkswagen presented an obvious test case of the new policy. In January 2016, the Department of Justice formally filed suit against Volkswagen. It was a civil complaint, but the allegations plainly suggested that criminal charges could follow. The complaint accused Volkswagen not only of deploying illegal software but also of orchestrating a cover-up beginning in 2014. "The United States' efforts to learn the truth" about the excess emissions, the Justice Department suit said, "were impeded and obstructed by material omissions and misleading information provided by VW entities."

And that was only one of Volkswagen's legal problems. U.S. law gave Volkswagen owners much broader scope to sue than owners in Europe. U.S. law allows owners to band together in class action suits and to seek punitive damages on top of the actual loss. Most European countries do not allow class action suits and do not provide for punitive damages. While Volkswagen admitted wrongdoing in the United States, the company argued that what it did was not illegal in Europe. EU emissions rules give carmakers broad leeway to dial back pollution controls in order to protect engines from damage, with no duty to inform regulators. It did not take long for personal injury lawyers in the United States to smell the opportunity. By mid-December 2015, at least five hundred lawsuits had been filed against Volkswagen. Besides owners, the suits came from used car dealers stuck with Volkswagen diesels on their lots, and even dealers for other car brands, who claimed that the illegal software gave Volkswagen an unfair competitive advantage. By spending less on emissions equipment, the suits said, Volkswagen was able to sell its cars for less than would otherwise have been the case. There were yet more suits from state governments claiming violations of their environmental laws and consumer fraud.

From Volkswagen's point of view, it was a legal nightmare. Plaintiffs were accusing Volkswagen of racketeering and seeking triple damages. They wanted the company to be compelled to buy back tainted vehicles at the original purchase price. Combined with the billions of dollars in fines that Volkswagen would conceivably have to pay federal and state authorities, it was not hard to imagine a worst-case scenario in which the total cost would add up to $50 billion or more. That sum might be enough to force Volkswagen into bankruptcy, endangering the jobs of Volkswagen's 600,000 workers worldwide, the vast majority of whom had done nothing wrong.

Volkswagen was also faring poorly in the court of public opinion. The company's claims that only a few rogue engineers were culpable inspired widespread ridicule. Felix Domke, a self-described hacker who lives in Lübeck, Germany, bought a used Bosch engine control unit on Ebay, the same EDC17 used in Volkswagen diesels, and was able to isolate the computer code that allowed the cars to detect when they were being tested. Domke, who had a day job in the IT department of a large corporation, which he preferred not to name, presented his conclusions in December 2015 to the Chaos Communication Congress, a hackers' convention in Hamburg. Using slides projected onto a large screen, Domke dissected how the engine software managed the SCR system in a Volkswagen. As expected, the car's computer took into account parameters such as engine speed, air temperature, and barometric pressure, as well as the amount of NOx being produced when deciding how much urea solution to inject into the pollution control system. Ideally, Domke explained, the car should inject just enough to neutralize the NOx emissions but not too much, or else the system would produce ammonia that would leak out of the tailpipe.

Oddly, though, the Volkswagen was programmed to inject far too little of the AdBlue urea solution almost all the time. Only under certain, very specific conditions, Domke said, did the computer provide enough of the solution to completely tame the NOx emissions. The car had to be traveling certain speeds for specific distances over an exact period of time. Those precise conditions would almost never occur in real life. The audience in Hamburg applauded when Domke projected on the large screen behind him a slide that showed how those speed, distance, and time parameters corresponded precisely with the simulated driving cycle used by regulators. "It's all based on detecting the driving cycle," Domke said. The computer code, he noted, was labeled "acoustic function."

The fact that a lone hacker could find the defeat device raised obvious questions about how Volkswagen could claim it did not. Domke appeared onstage with Daniel Lange, a former information technology executive at BMW with a firsthand understanding of how big auto companies function. Lange told the assembled hackers that there was no way the defeat device was the work of a small number of low-level employees. Software is heavily documented, and it is very unlikely that engineers would have taken the risk of modifying the code without approval at a high level of management, said Lange, who after leaving BMW founded a technology consulting firm called Faster IT. "It's completely unrealistic that a lonely engineer" would have been responsible, Lange said later.

Volkswagen executives came across as tone-deaf to the scale of public outrage, despite a shake-up in the Volkswagen press office and advice from numerous outside public relations firms. In January 2016, Matthias Müller made his first visit to the United States since becoming chief executive, attending the Detroit Auto Show. "We know we deeply disappointed our customers, the responsible government bodies, and the general public here in the U.S." Müller said at a news conference in Detroit. "I apologize for what went wrong at Volkswagen." He added, "We are totally committed to making things right."

So far so good. Then a reporter for National Public Radio approached Müller on the sidelines of the event. Müller had somehow become separated from his handlers and appeared unprepared for a spontaneous interview in his imperfect English. When the reporter referred to ethical problems at Volkswagen, Müller took umbrage. "It was a technical problem," he said. "An ethical problem? I cannot understand why you say that." Volkswagen had falsely interpreted U.S. law, explained Müller. "And we had some targets for our technical engineers, and they solved this problem and reached targets with some software solutions which haven't been compatible to the American law." Asked by the NPR reporter about the cover-up in 2014 and 2015, Müller replied, "We didn't lie. We didn't understand the question first. And then we worked since 2014 to solve the problem. And we did it together and it was a default of VW that it needed such a long time."

Müller's insistence that Volkswagen hadn't done anything unethical and didn't lie, despite ample evidence to the contrary, provoked an outcry and prompted him to backtrack the next day. "I have to apologize for yesterday evening because the situation was a little bit difficult for me to handle in front of all these colleagues of yours and everybody shouting," Müller said in a second interview with NPR. "First of all we fully accept the violation. There is no doubt about it. Second, we have to apologize on behalf of Volkswagen for that situation we have created in front of customers, in front of dealers and, of course, to the authorities," Müller said. "We had the wrong reaction when we got information year by year from the EPA and from the [California Air Resources Board]. We have to apologize for that, and we'll do our utmost to do things right for the future."

Volkswagen's relations with American authorities had not improved much, either. Citing Germany's strict data privacy laws, the company had refused to turn over internal e-mails and other documents that contained personal information about employees. "Our patience with Volkswagen is wearing thin," Eric T. Schneiderman, the attorney general of New York, said in January 2016. "Volkswagen's cooperation with the states' investigation has been spotty—and frankly, more of the kind one expects from a company in denial than one seeking to leave behind a culture of admitted deception."

Volkswagen had initially treated the emissions issue as a regulatory problem and was ill-prepared for a massive civil suit. The company had little experience with high-stakes U.S. litigation. As the situation deteriorated, in December 2015 the company replaced Michael Ganninger, its longtime general counsel, with Manfred Döss, who was general counsel for Porsche Automobil Holding SE, the Porsche-Piëch family holding company. For a company trying to convince the world it was raising its ethical standards, Döss was not necessarily the ideal choice. He had a reputation in Germany for playing rough, and he continued to serve as general counsel at the Porsche holding company, which raised potential conflicts of interest. But Döss had experience with U.S. litigation, and there was no doubt that he would ferociously defend Volkswagen's interests.

One of Döss's first moves was to name Robert Giuffra, a partner at the firm Sullivan & Cromwell, as national coordinating counsel in the U.S. litigation. Giuffra, based in New York City, already had a track record with Porsche and, by extension, with Volkswagen's largest shareholders. He had successfully defended Porsche in U.S. courts after a group of more than thirty hedge funds sued the company over the financial maneuvers it had deployed in its takeover bid for Volkswagen.

The funds, seeking to recoup the billions of dollars they had lost, hoped to take advantage of the United States' stricter securities laws. As previously noted, a federal appeals court ruled in 2014 that U.S. courts had no jurisdiction.

Besides winning that case, Giuffra had a blue-chip résumé. He earned an undergraduate degree from Princeton and a law degree from Yale, and worked as a clerk for William Rehnquist, chief justice of the U.S. Supreme Court. With a head of curly reddish hair, Giuffra also came across as direct and unpretentious. He was a native New Yorker and talked like one, sometimes sounding more like a taxi driver than a high-priced securities litigator at one of the city's most prestigious firms.

Giuffra and Döss were both fighters and averse to settling. But they were also realists. By admitting the existence of a defeat device in the United States, Volkswagen had already effectively pleaded guilty. Its only alternative was to negotiate the best possible settlement. Time was not on Volkswagen's side. The longer the tainted cars stayed on the road, the greater the excess pollution and the higher the potential fines. And the longer the court case dragged on, the greater the corrosion of Volkswagen's brand.

To represent Volkswagen top management, the company tapped Francisco Javier Garcia Sanz, a Spaniard on the company's management board. Garcia Sanz had been a protégé of José Ignacio López de Arriortúa, the General Motors purchasing whiz whom Ferdinand Piëch had poached from General Motors two decades earlier, prompting the scandal described earlier (see chapter 5). Garcia Sanz had been one of the team of so-called warriors who came with López from GM to Volkswagen. Garcia Sanz remained at Volkswagen after López was forced out, eventually becoming head of procurement, a powerful position that included control over supplier contracts worth tens of billions of euros.

Garcia Sanz had been on the Volkswagen management board since 2001, longer than anyone else. He oversaw the company-owned professional soccer team, VfL Wolfsburg, and hobnobbed with Spanish royalty. He possessed an affable charm, in contrast to the abruptness of some of his colleagues on the board. The fact that Garcia Sanz was not German, though he spent most of his career in Germany, may have provided a subtle advantage in dealing with Americans. He did not come across as a stereotypically stern Prussian. Garcia Sanz was also high enough in the organization that he could make decisions without constantly having to phone Wolfsburg.

Garcia Sanz and Döss moved quickly to reestablish a rapport with U.S. authorities. Beginning in January 2016, they regularly visited officials in the Justice Department, EPA, and CARB, promising to be more cooperative. Along with Giuffra, they worked to get around German data privacy laws and supply the U.S. investigators with the information they were demanding. Volkswagen employees in Germany were persuaded to waive their privacy rights so that documents could be transferred to the United States. In some cases, employees were allowed to delete personal information while a company lawyer watched.

WHILE VOLKSWAGEN GAVE GROUND in the United States, it took a hard line in Europe, refusing to admit it had broken any European laws. Oliver Schmidt, the same engineer who had taken part in the bungled negotiations with U.S. regulators in 2015, outlined the company's position in January when he testified before a committee of the British Parliament a few days after Müller's NPR interview. Schmidt, in a dark suit, with a shaved head and bags under his eyes, told members of the Transport Select Committee in the House of Commons that Volkswagen had removed the software that recognized when an emissions test was underway. But Schmidt called the software a "drive trace" and maintained, "This software is not defined as a defeat device in Europe."

Members of the committee, seated around a horseshoe-shaped table, were incredulous. "You are incorrect," said Louise Ellman, chairman of the select committee. "The German transport authority has been very clear in its ruling that what you were doing was not legal in Europe. Are you telling me that you have not removed that?"

"No," Schmidt replied, "I am telling you that we have removed it, but I said that in the understanding of the Volkswagen Group it is not a defeat device."

Volkswagen was trying to protect itself from legal liability in Europe, where the cost of having to pay restitution to 8.5 million diesel owners was potentially ruinous. The strategy might have made sense to a lawyer, but it conveyed the impression that Volkswagen was still in denial. And it risked insulting Volkswagen's European customers, many of whom had also bought the clean diesel sales pitch.

Meanwhile, the more conciliatory tone adopted by Giuffra and Döss in the United States began to pay off. In an effort to get a handle on the huge number of suits, a panel of federal judges ordered that all the Volkswagen cases be consolidated at a court in San Francisco overseen by U.S. District Judge Charles R. Breyer. The appointment of Breyer was one of Volkswagen's few lucky breaks. Breyer, seventy-four, a former aspiring actor and younger brother of U.S. Supreme Court Justice Stephen G. Breyer, had already handled several large, nationwide class action suits known as multidistrict litigation. He had a record of hacking through legal thickets and resolving cases quickly. One of his favorite sayings was "The perfect is the enemy of the good." Breyer made it clear from the first hearing, in January 2016, that his main concern was to fix the polluting cars or get them off the road, and to get owners the money they were due. He was not out to crucify Volkswagen or tar and feather its top executives. "It is obviously not a whodunit type of case," Breyer said at a preliminary hearing on January 21. "It is more of a case of how do we fix what was done? If it can't be fixed, then what is fair and just compensation for the people who have been damaged by this matter?"

Volkswagen began the talks in a weak position, though. U.S. regulators had rejected new plans to recall the offending vehicles. Since admitting the presence of defeat devices, Volkswagen had not been able to come up with changes to the engine software or other modifications, such as fiddling with the ignition timing or installing larger AdBlue tanks, that would make the cars legal without sacrificing gas mileage or performance. It was looking doubtful whether it was even technically possible to make all of the "clean diesels" live up to their hype. The only alternative was for the company to buy back and scrap all the cars, at an astronomical cost.

Hundreds of lawyers for owners and dealers came to the January hearing before Judge Breyer, so many that they filled the San Francisco courtroom to overflowing and spilled out into the hallway. All were eager for Breyer to appoint them to the plaintiffs' steering committee, which would take the lead in negotiations between owners and Volkswagen. Each applicant had a few minutes to convince Breyer that he or she belonged on the committee. Some of the United States' most prominent litigators lined up. They included Joseph F. Rice, a South Carolina lawyer who had negotiated a $206 billion settlement with the tobacco industry. Another was David Boies, the Washington lawyer who had represented Al Gore in the dispute over Florida ballots in the 2000 presidential election. Another was the former U.S. senator and presidential candidate John Edwards, returning to litigation after his political career had been derailed by a pregnant mistress. Edwards told the judge he had experience dealing with "foreign countries at the highest level."

Breyer selected the twenty-two members of the plaintiffs' steering committee before the day was out, acting with the same dispatch he had made clear he expected of the lawyers. Rice and Boies made the cut; Edwards and hundreds of others did not. Breyer appointed Elizabeth J. Cabraser, a veteran San Francisco product liability lawyer, as steering committee chairwoman.

Breyer also appointed Robert S. Mueller III, a former director of the FBI, as special settlement master. The job meant that Mueller was in charge of keeping the talks moving and acting as mediator. A partner at the WilmerHale law firm, Mueller kept a replica of a tommy gun on a credenza in his Washington office, where talks were often held. The weapon, a legacy of his FBI days, became a running joke in the months to come. When negotiations were at an impasse, somebody would suggest he put it to use.

Talks among the parties began in January and intensified the next month. Breyer imposed a tight schedule, saying he wanted to know by late March how Volkswagen was going to fix cars so that they would not be driving around spewing illegal amounts of harmful nitrogen oxides. The parties camped out in separate conference rooms at WilmerHale in Washington, subsisting on takeout food while Mueller shuttled among them, conveying offers and counteroffers.

The government, anxious to stem the excess pollution, also had an interest in a quick settlement. Giuffra's strategy was to focus on finding an agreement with the regulators, knowing that once the government had agreed on the outlines of a settlement, the lawyers for owners would have little choice but to go along. That led to some discontent among the plaintiffs' lawyers, who grumbled that they felt left out of the process. Now and then, they had to wait in rooms separate from where the talks were taking place, killing time.

There were tense moments, and occasions when people walked out in frustration. But all in all, participants said, the atmosphere was surprisingly collegial. "Everybody who was involved knew they were involved in something that was important," Giuffra said. "Some of the normal monkey business didn't happen."

In March, Volkswagen made a proposal. The company would fix the cars as best it could, while reducing emissions in other ways. For example, Volkswagen would pay to retrofit government trucks, buses, and even tugboats with cleaner engines. Some of the regulators were uneasy about the idea, which they felt would undermine clean air rules. But the worst-case alternative would be for states to refuse to register Volkswagen diesels, banning them from the roads and causing severe hardship for the owners.

On March 9, 2016, Garcia Sanz, Döss, Giuffra, and other members of the Volkswagen legal team trooped into the Department of Justice in Washington to hear whether the government would go along. The Volkswagen delegation cooled its heels in a conference room while government officials retreated to a private office belonging to John C. Cruden, an assistant attorney general representing the Justice

Department. There was some last-minute arguing about whether the government was conceding too much. "That was a big deal for us," said Mary Nichols, the chairwoman of CARB. Finally, the officials reappeared. They still wanted Volkswagen to offer buybacks to diesel owners. But the government representatives agreed that the company could give owners an option to keep their cars. If so, Volkswagen would upgrade the cars to make them as compliant with emissions standards as possible. Volkswagen accepted the compromise.

On April 21, after negotiating until 3 a.m. in Mueller's office the preceding Saturday, the parties met a deadline set by the judge to reach a preliminary agreement. It called for Volkswagen to buy back or fix diesels starting with the 2009 model year, to compensate owners, and to pay money to offset the environmental damage it had caused. But dollar amounts were left open, and there were still many issues to resolve, including technical details of upgrading the cars' emissions systems and the amounts that Volkswagen would compensate owners.

More intense negotiating sessions lay ahead. Breyer turned the grueling hours into a running joke. At a hearing on April 21 to discuss the preliminary agreement, the judge asked Giuffra how many hours he was putting in. When Giuffra estimated he had worked about four hundred hours the previous month, Breyer replied, "That's perfect." He added, "It's my expectation that even in this coming month you will be able to best your record again."

The financial and technical details were mind-numbing. The negotiators had to agree on how much the value of the vehicles had declined, and be able to adjust the figure according to the age and mileage of individual vehicles. They had to agree on what additional compensation owners should receive. They had to figure out what to do about people who had sold vehicles since September 2015, when the emissions fraud was exposed. They had to decide how to divvy up the money designated for environmental projects among the federal government and individual states. At one point, talks about how to value the cars bogged down over the meaning of "median" versus "mean."

Sharon Nelles, another Sullivan & Cromwell partner who worked closely with Giuffra, took a break in June 2016 to attend her daughter's high school graduation in New York. Two hours after the ceremony Nelles was on a plane back to Washington. When the private-practice lawyers like Nelles found themselves eating cold pizza at 3 a.m., they could at least take consolation that they were being paid by the hour. That was not the case for the salaried government negotiators, who included Cynthia Giles, an assistant administrator at the EPA responsible for enforcement, and Phillip Brooks, director of the EPA's Air Enforcement Division. On June 28, 2016, the parties announced an agreement and on June 30 presented it to Breyer. Giuffra was on the phone until the last minute getting officials from Texas—one of the states with the most Volkswagen diesels—to sign on. The agreement would cost Volkswagen a maximum of $14.7 billion. Of that amount, $10 billion was earmarked for owners. Those who opted to sell their cars back to Volkswagen would be entitled to receive a sum equal to the value of the car in September 2015, as estimated by the National Automobile Dealers Association. In addition, owners would receive restitution of at least $5,100 and as much as $10,000, depending on the value of the car. People who opted to keep their cars would receive the same restitution, as well as improvements to the emissions systems approved by the EPA and CARB. People with leased Volkswagens could turn them in without paying a penalty, and receive about half as much compensation as an owner would.

In addition, Volkswagen agreed to pay $2.7 billion into a trust fund that would be used to support programs designed to reduce nitrogen oxides in the atmosphere by at least as much as the tainted vehicles had generated in excess emissions. States and Native American tribes were eligible to apply for money they could use to replace old school buses or other vehicles with new, electrified vehicles; to equip municipal construction machinery with more fuel-efficient tires; or to undertake other projects that would lead to reduced pollution. Another $2 billion was earmarked for a program to be administered by Volkswagen that would promote battery-powered vehicles, including research and development and installation of public charging stations.

All sides could claim victory. The agreement allowed Volkswagen owners to get rid of their cars if they wanted to, plus get a decent sum as restitution. The settlement amount—much more than any previous emissions case—was large enough to serve as a deterrent to other automakers. Billions of dollars would flow to states.

The settlement with the United States in the civil case represented a major hit to Volkswagen's bottom line. But the damage could have been much worse. And it was possible that the final cost to the company of settling the class action would be lower than $15 billion. The $10 billion of that sum earmarked for owners was based on the cost of buying back all 500,000 diesel Volkswagens and Audi A3s. If some owners chose the option to keep their cars, which seemed likely, the cost to Volkswagen would be much less. In addition, along with other carmakers, Volkswagen stood to benefit from the investments in zero-emissions technology, which would help create a market for battery-powered cars. Under the agreement, Volkswagen retained control over the $2 billion for electric vehicle projects, though the company was obligated to spend the money in a way that would benefit all manufacturers of battery-powered cars.

Perhaps most importantly, the settlement removed a huge source of financial uncertainty for Volkswagen. The company had postponed publishing its annual report for 2015 while the settlement talks were underway, in part because the company's outside auditor, PricewaterhouseCoopers, refused to endorse the financial data without a more definite idea of what the scandal would cost. The delay provoked a series of other problems. Without an annual report, Volkswagen had to reschedule its annual shareholders meeting. Volkswagen Financial Services temporarily stopped issuing bonds, because it could not provide investors the financial information they required. When Volkswagen finally did report earnings, several weeks before the settlement was announced, the loss of €1.6 billion, or $1.8 billion, was the first annual loss since 1993, the beginning of the Piëch era.

But the U.S. settlement was by no means the end of Volkswagen's legal troubles. The agreement did not cover the eighty thousand Audi,

Porsche, and Volkswagen cars with three-liter motors. Volkswagen was still trying to find a way to fix the larger cars that would be acceptable to the EPA and CARB. The agreement also did not include fines that might still be imposed by the federal government as well as by individual states like New York, Maryland, Massachusetts, Pennsylvania, and Vermont that had filed separate suits. Nor did the agreement have any effect on the Justice Department's criminal investigation.

AT THE SAME TIME that the class action case was occupying U.S. courts, the circumstances surrounding Porsche's attempted takeover of Volkswagen years earlier resurfaced in the German justice system. Prosecutors in Stuttgart had slowly but doggedly gathered evidence that top Porsche executives had resorted to illegal means after the advent of the financial crisis in 2008 threatened to unravel the complicated derivatives play they used to get control of Volkswagen shares. First the prosecutors charged Härter, the former chief financial officer, with credit fraud for attempts in 2009 to persuade the French bank BNP Paribas to provide financing to continue the takeover bid. Härter was convicted in June 2013 and paid a €630,000 fine ($880,000) after his appeals failed. A subordinate was also convicted of credit fraud and fined, but nevertheless given a job afterward by Volkswagen—as a financial controller. By Volkswagen standards, anyway, the man could still be trusted with company money. (According to a Porsche spokesman, the company regards the subordinate as a victim of circumstances. Because he had no prior criminal record, he was given a second chance.)

Building a case against Wiedeking, the former Porsche chief executive, was harder. The state court in Stuttgart initially dismissed an indictment against him and Härter for alleged violations of securities laws, but was ordered by an appeals court to take the case to trial. The trial, which began in October 2015, took place at the state court in Stuttgart, a boxy eight-story building with a façade of beige brick that was built in the 1950s after the original building was destroyed by bombs in World War II. A plaque memorializes people executed by the

Nazi regime in what is now an interior courtyard. On days when court was in session, Wiedeking arrived in a black, chauffeur-driven Panamera, the luxury four-door sedan that Porsche began selling in late 2009. The car was developed on Wiedeking's watch, and was a further example of VW-Porsche cooperation. The chassis were produced at a Volkswagen factory in Hanover, then finished off alongside the Cayenne at the Porsche factory in Leipzig.

The central question was whether the two executives deliberately pumped up Porsche's share price by creating the false impression that there was a shortage of shares on the market, provoking panic among the speculators and a short squeeze. In an opening statement to the court, Wiedeking was indignant. He accused prosecutors of being tools of the hedge funds. He denied that Maple Bank had been on the verge of withdrawing its support for the Porsche options strategy. He insisted that Porsche had always communicated accurately. "I have nothing to be ashamed of," Wiedeking told the court.

For the two young prosecutors representing the state, Heiko Wagenpfeil and Aniello Ambrosio, it was lonely work. The presiding judge, Frank Maurer, was openly skeptical of their case. The German news media generally gave them little chance of success. Wiedeking and Härter were represented by legal teams that boasted the best of the German defense bar. Hanns W. Feigen, Wiedeking's lead defender, was one of Germany's go-to lawyers for prominent business executives in danger of going to jail. Härter was represented by Sven Thomas, whose clients had included Bernie Ecclestone, the billionaire chief of Formula One racing. With Thomas's help, Ecclestone avoided incarceration on a German bribery charge by agreeing in 2014 to pay a €100 million ($135 million) fine.

The prosecution in the Wiedeking and Härter trial was handicapped by a lack of witnesses who would testify that Porsche had been less than honest with investors about its plans for Volkswagen. Some of the former Porsche bankers suffered memory losses on the witness stand. Executives from Maple Bank said it wasn't true that Porsche was about to run out of money. In the chaos that followed the collapse of Lehman

Brothers, the defense lawyers argued, it was impossible to demonstrate that the press release Porsche issued in October 2008 was responsible for the violent fluctuations in Volkswagen's share price that followed.

Still, there was some suspense in the weeks leading up to March 18, 2016, when Judge Maurer was scheduled to deliver the verdict by a five-judge panel. Wagenpfeil and Ambrosio, the state's attorneys, who had sometimes seemed bumbling during the trial, delivered surprisingly effective final arguments. Acknowledging that they had not been able to prove much of their case, they focused on the October 24, 2008, press release disclosing that Porsche had locked up 74 percent of Volkswagen shares. They argued, in effect, that there was no other plausible explanation for the statement, which provoked scramble by short sellers to buy shares, except as a desperate attempt to prop up Volkswagen's stock price. Certainly Wiedeking and Härter weren't, as the press release claimed, trying to be considerate to the hedge funds.

The courtroom was crowded with reporters on the day of the verdict. Härter arrived first with his lawyers, followed by Wiedeking in a dark suit and rimless glasses. "Guten Morgen," Wiedeking said to the press, with a hint of irony in his voice. Then he took a seat with his lawyer in the front of the courtroom. Lawyers associated with Volkswagen and the Porsche and Piëch families arrived and sat with the defendants. Among them was Döss, who retained his post as chief counsel for Porsche SE, the company that holds Porsche and Piëch families' VW shares. In his role as lawyer for the holding company, Döss had an interest in the verdict. But it was odd for the chief counsel of Volkswagen to be taking sides in a criminal trial by sitting among the defense lawyers. Also present was Matthias Prinz, a lawyer for Ferdinand Piëch. No members of the Porsche and Piëch families were charged in the case, but they had a financial stake. If Wiedeking and Härter were convicted, Porsche SE would have to give up €800 million ($880 million) in what prosecutors said were ill-gotten profits. All told, there were eleven defense lawyers. The two prosecutors, Wagenpfeil and Ambrosio, sat alone.

Judge Maurer took the bench and quickly delivered the verdict: not guilty on all counts. Then, for the next two hours, he lectured Wagenpfeil and Ambrosio on what he believed were the inconsistencies and gaps in their case. Among other things, the judge said, he was not convinced there had even been a shortage of Volkswagen shares in October 2008. Amid the market tumult then rampant, there may have well been some other "irrational factors," Maurer said, to explain the surge in the share price that made VW briefly the most valuable company in the world. The press release that Porsche issued just before the spike in the share price did not contain any market-moving news, the judge maintained. While it was clear that Porsche "was not Mother Teresa," he said, the company had been open about its intentions toward VW. "There was no secret plan to take over Volkswagen," Maurer concluded.

It was a public humiliation for the two prosecutors, who sat at tables with dark veneer surfaces and chrome legs, surrounded by empty chairs, their backs literally against a courtroom wall. When the judge's two-hour explanation of the decision was over, the news media and the lawyers spilled out into the courtroom lobby, where Härter said that the prosecutors should face consequences for unfairly persecuting him.

The prosecutors appealed the verdict, but later withdrew the appeal, citing slim chances of success. Once again the Porsche and Piëch families had survived a crisis with no lasting consequences. The only possible damage was that the trial called attention to the circumstances surrounding the Volkswagen-Porsche merger and the degree to which the transaction had served to enhance the families' power and wealth. For those who put some stock in prosecutors' version of events, the trial illuminated how close the family had come to financial ruin. But instead, the Porsches and the Piëchs managed to emerge with de facto control of Volkswagen for the first time since World War II.

Representatives of the Porsche family insisted that their financial circumstances were never as perilous as the state's attorneys claimed, and even argued that the family had been better off owning a profitable sports car maker rather than Europe's largest carmaker. It was

Ferdinand Piëch himself, with his sometimes startling candor, who had already undermined the assertion that the family fortune was never in danger. "The first generation builds," Piëch told a television interviewer. "The second maintains. My generation is the third. We normally destroy. In 2008, we almost did."

BACK IN WOLFSBURG, Volkswagen continued to insist that it was changing its ways. Effective at the beginning of 2016, the company had hired Christine Hohmann-Dennhardt, a former justice on Germany's highest court, as a member of the management board responsible for integrity and legal affairs. She became the first female member of the Volkswagen management board in history, and the first board-level head of compliance. Her job was to upgrade Volkswagen's obviously inadequate systems for making sure employees did not break any laws. Müller personally apologized to President Barack Obama for Volkswagen's behavior when the two met briefly at a dinner in April hosted by Chancellor Angela Merkel in Hanover.

But some of Volkswagen's actions raised doubts about how much had really changed. Despite the record €1.6 billion loss for 2015, the Volkswagen supervisory board initially refused to cut bonuses for the twelve men who had been members of the management board during 2015. They were scheduled to receive a combined €35 million ($38.5 million) in performance-related pay. The board relented after a public outcry. The cut of €4.2 million ($4.6 million) from the total was only a fraction of what the executives were supposed to receive. And they stood to recoup the cuts and then some if Volkswagen shares recovered by 2019. The members of the management board still received total compensation of €59 million ($65 million) among them, including €31 million ($34 million) in bonuses. Müller, for example, received €3.9 million ($4.3 million) in salary and bonuses, rather than the €4.8 million ($5.2 million) he would have received without a cut. Even Winterkorn, on whose watch the fraud and cover-up had taken place, received salary and bonuses of €7.3 million ($8 million) for the less than ten months of

2015 he worked before resigning. Unlike the others, Winterkorn was not forced to take a cut in his bonus.

The members of the supervisory and management boards got an earful when the annual meeting finally convened on June 22, 2016, more than a month late, at a vast exhibition hall at the Hanover fairgrounds. "They have been rewarded for failure," said Hans-Christoph Hirt, a director of Hermes EOS, an organization that advocates on behalf of large shareholders. Hirt told members of the supervisory board—dominated by the Porsche and Piëch families and organized labor—that they were ultimately responsible for a "culture in which the emissions scandal was able to unfold and remain undetected for many years."

The board members listened impassively from a stage. With less than 11 percent of Volkswagen's voting shares in the hands of outsiders, they could easily ignore such criticism. The families as well as the state of Lower Saxony stood behind management. So did the company's other major shareholder, the sovereign wealth fund of Qatar. Together, they crushed a motion by dissident shareholders to remove Pötsch as chairman of the supervisory board on the grounds that he had been chief financial officer when the wrongdoing was taking place. Speaking at the meeting, Müller continued to give the impression that the fraud was perpetrated by a small number of employees. "Based on what we now know, in the past there were certain process deficits in some technical subdivisions in addition to misconduct on the part of individuals," he said. Pötsch denied that top executives were involved in a cover-up, saying they were not aware of the gravity of the emissions problem until shortly before the EPA issued a formal complaint.

Yet the pressure from outside was relentless. Two days before the shareholders' meeting, prosecutors in Braunschweig disclosed that they were investigating whether Winterkorn and Herbert Diess, the head of the Volkswagen brand, had violated German securities laws by failing to notify shareholders sooner of the emissions problem. The investigation was based on an inquiry by BaFin, the German banking and stock market regulator. In fact, the agency had requested that all of the members of the Volkswagen management board in 2015 be held collectively responsible for illegally keeping shareholders in the dark. BaFin officials were annoyed that prosecutors pursued only Winterkorn and Diess. Investors in Europe and the United States sought damages for the decline in Volkswagen shares they had suffered. The potential judgment could be in the billions.

Customers in Europe were also restive. After the civil settlement in the United States, which was worth about $20,000 per car, European owners wondered why they were supposed to be satisfied with a software update and, in some cases, a length of plastic tubing. One of the disgruntled owners was Jürgen Franz, a retired advertising executive in Munich. "Why are they getting so much and we're getting nothing?" Franz asked. He drove his diesel Tiguan SUV the same twelve-kilometer route every morning to a lakeside park to go running with his dog. After Franz brought the car in for the software update—which took fifteen minutes, he said—fuel economy suffered noticeably. Franz said he had received a call from a consumer survey firm asking whether he would buy a Volkswagen again. "I said 'no,'" Franz declared.

The risks for any Europeans like Franz who might contemplate suing Volkswagen or any deep-pocketed corporation are substantial. European countries typically require the losing party in a suit to pay the legal costs of the winner. Countries like Germany do not allow lawyers to work on a contingency basis, collecting a percentage of the winnings if the suit is successful and nothing if it's not, as is common in the United States. Any Golf or Tiguan owner in Europe who sued Volkswagen would incur the risk of having to pay Volkswagen's lawyers if the suit failed.

But people in Europe were finding creative ways to get around these obstacles. Those who searched Google for the German words for "Volkswagen damage claim" would see an ad for the website My-right.de, one of several start-ups using the Internet to recruit Volkswagen owners. In Paris, Weclaim.com was attracting French customers. The firms took advantage of a loophole that allows European consumers to sign their legal claims over to third-party service providers, which then try to recover damages, collecting a commission if they do. The model, risk-free for consumers, had been used in the past to allow large numbers of people to pursue relatively small claims against airlines, banks, or other corporations. The Internet made it possible to recruit huge numbers of consumers who shared similar gripes, and to automate the processing of their claims, creating de facto class action suits. Volkswagen seemed almost certain to be the biggest such case ever. The websites had teamed up with lawyers and were expanding into other European countries. The European claimants were seeking a mere €5,000, or about $5,500, per car. But even if they collected only half of that, the sums multiplied by millions of European customers would add more billions to the financial burden on Volkswagen.

Volkswagen's legal exposure in the rest of the world was less severe because it had never sold many diesels outside of Europe or North America. There were fewer than 2,000 Volkswagen diesel cars in China, for example. Still, China recalled the vehicles for repairs. In Brazil, authorities fined Volkswagen the equivalent of $13.2 million, the maximum under Brazilian law, for illegal software in about 17,000 Amorak pickups, the only diesels that Volkswagen sold in the country. There were also investigations or mandatory recalls in numerous other countries, including South Africa, Australia, and India. South Korea, one of the few countries in Asia where diesel cars are popular, took particularly harsh measures. Government authorities raided Volkswagen offices, indicted a local Volkswagen official for falsifying documents, and banned sales of eighty Volkswagen diesel and gasoline models, all but driving the company from the market. South Korea also ordered a recall of 126,000 cars and imposed fines of nearly $30 million. But other countries were reluctant to confront Volkswagen. In Canada, Volkswagen owners complained about government inaction even though some 100,000 tainted diesels were on the country's roads.

In the United States, authorities signaled that the civil settlement in Federal court had not dimmed their determination to hold individual Volkswagen executives accountable. The suits filed by New York, Massachusetts, Maryland, Vermont, and other states, which claimed violations of consumer protection and pollution laws, ruptured the secrecy surrounding Volkswagen's internal investigation. By mid-2016, investigators had been able to comb through large numbers of e-mails and documents to construct a much clearer picture of what took place inside Volkswagen, and who was involved. According to the state suits, Volkswagen managers routinely resorted to defeat devices when they ran into technical roadblocks.

Complaints filed by the states of New York, Maryland, and Massachusetts alleged that Volkswagen had installed six separate defeat devices over more than a decade. The first was Audi's "acoustic function," beginning with 2004 models, which shut down emissions systems when three-liter diesel motors were starting up. The alleged purpose was to reduce the noise the motors made when cold. The second defeat device was an adaptation of the acoustic function to allow the 2009 model Volkswagen and Audi diesels with two-liter motors to pass U.S. emissions tests. The third was installed in Audis with three-liter motors, beginning with 2009 models. That defeat device compensated for the decision by Audi not to give the cars a big enough urea tank to meet U.S. standards. The alleged fourth defeat device was installed in Passats beginning with 2012 models, again because they did not have large enough urea tanks. The inadequate tank was also the reason for the fifth defeat device, supposedly installed in Porsche Cayenne SUVs beginning with 2013 models, the first Porsche diesels to be sold in the United States. The sixth defeat device was installed in 2015 Passats, Jettas, Golfs, Beetles, and Audi A3s, all of which had SCR systems with undersized urea tanks. The use of defeat devices was common knowledge inside Volkswagen, the suits alleged.

In contrast to the prosecutors in Braunschweig, who were restrained by strict German privacy rules, the U.S. investigators were not shy about naming names, or of accusing top managers of being complicit in the wrongdoing. Complaints by New York, Maryland, and Massachusetts were the first to allege that Winterkorn and a person identified in Volkswagen documents as "H. Müller" were aware as early as 2006 that the urea tanks in Audis with three-liter motors were not big enough to meet U.S. emissions standards, the complaints said. Rather than redesign the cars to accommodate larger tanks, Volkswagen decided to deploy software that would thwart emissions tests, the suits alleged. Company spokespeople reacted indignantly to questions from reporters about whether the new chief executive should suffer any consequences.

The U.S. investigators were not impressed by Volkswagen's promises to reform. The bonuses that top management received despite the record loss and plunging share price, the New York complaint said, "demonstrate that the company's culture that incentivizes cheating and denies accountability comes from the very top and, even now, remains unchecked."

Volkswagen, clearly peeved that the $15 billion settlement had not put an end to the company's legal troubles in the United States, was in for a bigger shock. On September 9, 2016, James R. Liang, a sixty-two-year-old engineer who had worked at Volkswagen since 1983, appeared before a federal judge in Detroit and pleaded guilty to charges that included fraud and conspiracy to violate the Clean Air Act. Originally from Singapore, Liang had been part of the team in Wolfsburg that designed the EA 189 motor. When the clean diesel push began in the United States, Liang moved to the Volkswagen testing facility in Oxnard, near Los Angeles, where his duties included supervising testing and certification of diesels for sale in the United States. In 2014 and 2015, after the West Virginia University study prompted CARB to start asking questions, Liang was part of the group at Volkswagen that concocted bogus explanations for why the diesels polluted so much outside the testing lab, according to the indictment against him. In short, Liang, who continued to live in California, had been in the thick of things since the beginning. Under a plea agreement, if he cooperated to the satisfaction of the government, he would receive a reduced sentence.

As a midlevel engineer, Liang was a relatively small player in the conspiracy. But with his broad knowledge he could be used to put pressure on others, including Volkswagen employees in Germany.

Germany does not normally extradite its own citizens. But if U.S. prosecutors decided to indict German executives, those people would have to be very careful about which other countries they visited for the rest of their lives. They would be safe only in Germany and the limited number of countries, mostly in the developing world, that do not have extradition treaties with the United States. The list includes Russia, China, and Saudi Arabia. Other European countries offered no refuge. Only three years earlier, FBI agents working with Italian authorities had arrested Florian Homm, a German hedge fund manager wanted on securities fraud charges, in a Florence art gallery. (Homm was released from an Italian jail before he could be extradited. He hurried back to Germany, where he was safe from United States authorities. He was never tried.) According to the doctrine presented in the Yates memo—the 2015 appeal by a top U.S. official for more accountability by top corporate executives—the Justice Department would pursue charges as high up in the Volkswagen organization as the evidence would go. "The conduct and the facts and the seriousness of the violations warrant reaching well beyond our shores," said William Sorrell, the attorney general of Vermont. Speaking in his office in the state capitol in Montpelier a few days after Liang's guilty plea, Sorrell said, "My sense is that a number of people haven't slept quite as well since."

CHAPTER 21

Punishment

"We are interrupting this broadcast," intones an announcer, "for an official threat to Germany from the United States of America." So begins a satirical video broadcast in November 2015 by ZDF, one of Germany's main television networks. "Manipulated Volkswagens with counterfeit emissions readings have destroyed our beautiful environment," the narrator says. "Therefore in the future we will only buy vehicles produced in America." Cut to shots of U.S.-made monster pickups spewing clouds of black smoke. Hard rock music pounds in the background as the narrator recites statistics about the trucks' massive horsepower and dismal fuel economy. There are shots of overcrowded American freeways and more big trucks interspersed with shots of women in bikinis firing automatic weapons—apparently a comment on American gun culture. The final scene shows the Stars and Stripes billowing in the wind. "American autos," the narrator says sarcastically. "Not manipulated. For the love of the environment."

The video expressed what many Germans see as the hypocrisy behind the legal onslaught against Volkswagen in the United States. How, they ask, can Volkswagen, a maker of fuel-efficient cars, suffer such severe penalties in a country that uses more energy per person—about 80 percent more—than Germany? Why should Volkswagen pay much more than other auto companies that have also polluted or whose flawed products in some cases caused people to die?

On September 17, 2015, only a day before the EPA cited Volkswagen for emissions cheating, General Motors agreed to pay a $900 million fine to settle charges stemming from defective ignition switches. The switches, installed in Chevrolet, Pontiac, and Saturn models from 2005 to 2007, sometimes turned off while cars were underway, disabling the power steering and brakes as well as the air bags. The defect, which some engineers inside GM knew about but failed to disclose, was linked to the deaths of 124 people. In 2005, the company rejected a fix that would have cost less than $1 per car. GM dismissed fifteen employees as part of the 2015 settlement, but the Justice Department did not pursue criminal charges against any individuals. All told, recalls, fines, legal fees, and compensation to the families of victims cost GM somewhat over $6 billion, based on information in the company's 2015 annual report. In other words, a product flaw that killed more than 100 people cost GM less than half of what Volkswagen had to pay just to compensate owners in the United States and atone for the environmental damage. Germans understandably ask whether that is fair.

Volkswagen is also by no means the only company making cars that spew excess amounts of nitrogen oxide pollution. After the emissions scandal came to light, European governments began to scrutinize other manufacturers more closely. Studies by Germany and others, which included road testing, did not turn up proof that other carmakers went as far as Volkswagen, installing defeat devices to conceal illegally high emissions. But almost all other manufacturers took advantage of loopholes in European regulations allowing them to reduce emissions controls in order to protect engines under certain conditions, such as colder weather. As a result, European cities are much more polluted and unhealthy than they should be.

For example, according to a study by the German government, a diesel Jeep Cherokee, sold in Europe by Fiat Chrysler, throttled down the emissions controls when the outside air temperature was below twenty degrees Celsius, or about sixty-eight degrees Fahrenheit—not exactly arctic conditions. In some conditions, the study reported, the Jeep's nitrogen oxide emissions were twelve times as high as allowed. (Fiat Chrysler declined to comment about the German government's findings.) Although the study found that the Jeep was among the worst polluters, it also found that cars by practically all the manufacturers in Europe, including GM's Opel unit, Mercedes-Benz, Renault, and BMW, exceeded pollution standards in everyday use. (GM, Mercedes, and other European manufacturers replied to the study report by saying they were working on improving their emissions systems.) A British government study of about forty cars found that they spewed an average of six times the legal limit of nitrogen oxides under normal driving conditions. In some cases, Volkswagens with illegal software polluted less than competitors' vehicles that legally took advantage of lax European Union rules. The government reports have fed resentment in Germany that Volkswagen is being unfairly singled out. Everyone cheated, the argument goes. Sure, Volkswagen was the only one to deploy defeat devices. But the outcome was basically the same.

So why is Volkswagen being made to pay so much?

Volkswagen's defenders argue that the amount of nitrogen oxides from the 600,000 VW and Audi diesels in the United States was a drop in the bucket compared with the amount pumped into the atmosphere by long-haul trucks, power plants, and other polluters. The excess pollution from the Volkswagens amounted to less than 0.001 percent of the total nitrogen dioxide generated by manmade sources. Still, it was enough to do substantial damage. According to a study published in September 2016 in the *International Journal of Environmental Research and Public Health*, the tainted Volkswagens produced anywhere from about 3,400 to 15,000 more metric tons of nitrogen oxides per year than they should have. (The study was conducted by a team from Northwestern University, the University of Texas, Columbia University, and Harvard.)

One of the great and probably underappreciated achievements of the second half of the twentieth century was a drastic reduction in air pollution. Emissions control technologies, mandated by stricter regulations, have cut levels of nitrogen dioxide, the most harmful member of the nitrogen oxide family, by almost half in urban areas since 1990. With its defeat devices, Volkswagen evaded those restrictions and undermined the progress made. Despite the stereotypes portrayed in the satirical video by ZDF, the Volkswagen diesels tested by West Virginia University produced far more nitrogen oxides than a full-size, late-model American diesel pickup truck. (If a truck spews clouds of black smoke, as in the ZDF video, it is probably because the owner has illegally disabled the emissions system.) In fact, according to CARB, a 2010 Jetta diesel sold in the United States produced about one gram per mile of nitrogen oxides, twice as much as a long-haul diesel truck from the same model year.

As a result of the excess nitrogen dioxide produced by the Volkswagens, anywhere from 5 to 50 people would suffer premature deaths from respiratory ailments, according to the study in the *International Journal of Environmental Research and Public Health*. Another 250 to 1,000 people would suffer respiratory ailments, including asthma attacks or acute bronchitis. A separate study by researchers at the Massachusetts Institute of Technology and Harvard University came to similar conclusions. About 60 people will die ten to twenty years earlier than they would have otherwise because of the Volkswagens, and the number will grow the longer the vehicles stay on the road.

Such studies have flaws. It is impossible to link the health effects of Volkswagen emissions directly to individual deaths and illnesses. There is no way to follow nitrogen dioxide molecules from a Jetta tailpipe into the lungs of someone who then gets sick. The researchers in the academic studies lacked data about where the offending Volkswagens are located and how many are in heavily populated urban areas where they are more likely to cause health problems. The cause and effect is not as immediate and devastating as an ignition switch that disables a car's systems on a superhighway, leading to a fatal crash. Still, it is possible to say with reasonable certainty that the pollution caused by Volkswagens led some people to get sick who wouldn't have otherwise, and in some cases to die an early death.

In the United States, Volkswagen bears most of the blame for pollution originating from diesel passenger cars. No other car company has marketed diesel as intensively as Volkswagen. Sales of diesel cars in the United States were minuscule before Volkswagen began the clean diesel push, and there is no evidence that the few other companies that sell diesels in the United States committed similar misconduct. (The BMW tested by West Virginia University performed well.)

In Europe, it's true that Volkswagen is not the only car company responsible for abysmal air quality in major cities, where nitrogen oxide levels have remained high even though pollution regulations have theoretically become stricter. On paper, carmakers have cut nitrogen oxide emissions by passenger cars by 80 percent since 2000. But those figures are based on laboratory tests. Actual emissions under normal driving conditions have fallen only 40 percent.

In Europe, governments share the blame for excessive pollution from diesels. Enforcement of pollution standards among the twenty-eight members of the European Union is inconsistent or nonexistent. It is very unlikely that European regulators would have ever discovered the Volkswagen defeat devices without the work of the International Council on Clean Transportation, West Virginia University, and CARB. EU rules do not define defeat devices clearly, leaving plenty of room for automakers to evade emissions rules without breaking the letter of the law. When Jeep engine software dialed back emissions equipment below twenty degrees Celsius, the company could say it was protecting the motor. But it was also the case that the EU test procedure was conducted at temperatures above twenty degrees. Jeep engineers could be sure that the regulators would not register the excess emissions, at least not until Volkswagen called attention to the issue. In that sense, what Jeep and other carmakers did was the functional equivalent of a defeat device. By flouting the spirit, if not the letter, of the law, the carmakers undermined efforts by regulators to clean up the air. European governments were enablers because they did not enforce the rules.

Still, there is a distinction between what Volkswagen did and what its rivals did. An analogy would be a company that uses every possible loophole to reduce its taxes, and another that simply lies to tax authorities. The first company might be unpatriotic and reprehensible, but it would not be breaking the law. The second company, even if it paid more taxes, would be guilty of a criminal offense. Volkswagen's willingness to resort to illegal methods, when others (so far as is known) at least followed the letter of the law, is a reflection of the company's culture.

As the largest carmaker in Europe, Volkswagen bears a disproportionate share of blame for alarmingly high levels of nitrogen oxides in almost all European cities. Volkswagen, more than any other carmaker, was responsible for the proliferation of diesel passenger cars in Europe since the early 1990s. Under Ferdinand Piëch, Volkswagen and the Audi division pioneered diesel in passenger cars and were proud of it. By using defeat devices, Volkswagen cut the cost of its emissions equipment and put enormous competitive pressure on rivals. It would have been difficult for other European car companies to compete with Volkswagen, with its 25 percent market share, unless they also cut corners. As Europe's biggest car company by far, Volkswagen had a special responsibility to set the standard. Instead, it established a lowest common denominator.

When Volkswagen set out to popularize diesel in the United States, the company's executives behaved as if they were in Europe with the same lax rules, market dominance, and political clout. It is hard to say whether company executives were ignorant of American laws or so arrogant they did not feel bound by them. Defeat devices are well defined by EPA regulations. The agency provides extensive guidance to automakers about what is and what is not allowed. Such guidance is essential because of the complexity of modern automobiles and the reliance on engine computers to constantly adjust variables such as the ignition timing or the volume of exhaust gas recirculation. It is not always easy to demarcate the boundary between normal operations and cheating. "In Europe," says John German of the International Council on Clean Transportation, "you have a pretty good definition of what a defeat device is, but there is a lack of guidance that creates this huge gray undefined area. In the U.S. it's not gray, it's not undefined. It's all laid out."

Volkswagen managers and engineers seem to have been unaware of the work done by Leo Breton at the EPA and companies like Horiba to make road testing of vehicle emissions feasible, thereby increasing the risk of getting caught. And when Volkswagen was caught, its leaders underestimated the wrath of U.S. regulators. "Every other manufacturer was smart enough to remove defeat devices. VW was the only one stupid enough" not to, German said.

Having used deceit to meet the United States' stricter limits on nitrogen oxides, Volkswagen doubled down by aggressively promoting its cars as environmentally virtuous. That strategy elevated the emissions cheating from a mere violation of regulations to a gigantic consumer fraud. GM had to compensate the families of people who died because of its faulty ignition switches, as well as those who were injured. But GM owed nothing to customers who had not been in accidents, limiting its financial liability. Volkswagen, with its flagrantly deceptive advertising, exposed itself to claims from every single owner of a diesel vehicle in the United States, as well as to suits by states claiming that Volkswagen violated their consumer protection laws. The settlement in Judge Breyer's court was relatively modest on a per car basis, about $20,000. But because there were so many cars, the total cost to Volkswagen dwarfed what GM paid to victims of faulty ignition switches.

Volkswagen further inflated its financial liability by failing to cooperate fully with CARB and the EPA after the West Virginia study raised the first suspicions. GM's behavior in the case of the defective ignition switches provides an illuminating contrast. After learning of the problem, Mary Barra, the GM chief executive, appeared before a subcommittee of Congress and promised total openness with investigators. "Whatever mistakes were made in the past, we will not shirk from our responsibilities now and in the future," Barra said on April 1, 2014. "Today's GM will do the right thing." GM then delivered. The company appointed a former federal prosecutor, Anton Valukas, to investigate why GM did not recall the defective cars until 2014 when it had known about the problem at least since 2005. Valukas "has free rein to go where the facts take him, regardless of the outcome," Barra told the congressional committee. "The facts will be the facts."

In the view of the U.S. Justice Department, Barra's promise was not just lip service. GM conducted "a swift and robust internal investigation," the department said in 2015, providing federal agents "with a continuous flow of unvarnished facts." GM also provided "without prompting, certain documents and information otherwise protected by the attorney-client privilege." The company had shown "acceptance and acknowledgement of responsibility for its conduct," the Justice Department said. GM's behavior helped lower the fine the company had to pay.

Volkswagen bobbed and weaved for more than a year before admitting it had used defeat devices. By then it had used up regulators' patience and could expect no mercy. Not only consumers were duped by the "clean diesel" pitch. "We've had a focus on diesels for a long time and frankly skepticism about the whole clean diesel program," said Mary Nichols, the chairwoman of CARB. "But our engineers were convinced by the German engineers, Volkswagen certainly being one of the leaders, that they had found a way to meet our standards and still do it with vehicles that are very, very fuel efficient."

"Many very smart people bought these vehicles because they thought it was a great breakthrough," Nichols said. "So they were even madder than people like me when they found out that they had been deceived."

Even after its confession, Volkswagen continued to dig a deeper financial hole for itself by moving slowly to clarify how the cheating had come about and by failing to hold managers accountable. The evidence presented in the state lawsuits and the indictment of the Volkswagen engineer James Liang left no doubt that U.S. authorities viewed the emissions fraud as part of a broad conspiracy. While several dozen midlevel engineers and executives had been suspended by late 2016, the supervisory board did not discipline any members of the management board. Even assuming, for the purposes of argument, that top managers like Matthias Müller or Rupert Stadler, the head of the Audi division, didn't know about the defeat devices until Volkswagen admitted their existence, the question remains: why didn't they know? Wrongdoing serious enough to threaten the existence of the company was taking place under their noses. Where were the checks on employee behavior that would have prevented such behavior? Members of the management board also bore responsibility for Volkswagen's continued foot-dragging even after the company admitted existence of the defeat devices. Yet they kept their jobs and received their bonuses, minus a token sum.

Siemens, another iconic German company, could have provided a useful model for Volkswagen. The electronics and engineering giant, based in Munich, was accused in late 2006 of bribing foreign officials on a grand scale in order to win contracts to build projects like urban rail networks in Venezuela or a mobile phone system in Bangladesh. Like those at Volkswagen, lower-level executives were under immense pressure to meet targets, and breaking the rules was tolerated. Siemens' initial response was to downplay the allegations. But within six months the chairman of the supervisory board, Heinrich von Pierer, and the chief executive, Klaus Kleinfeld, resigned under pressure and were replaced by company outsiders. The new chief executive, Peter Löscher, hired a well-known anticorruption expert to oversee reforms, increased the number of internal compliance staff to five hundred, from fewer than one hundred, and disciplined nine hundred employees, many of whom were dismissed. The measures helped demonstrate to prosecutors that Siemens was serious about change. "Siemens' cooperation, in a word, has been exceptional," Assistant U.S. Attorney General Matthew Friedrich told reporters on December 15, 2008. "Siemens has faced facts, accepted responsibility, retained experienced counsel to conduct thorough internal investigations, and has implemented real reforms." Siemens, which had a New York stock listing and large operations in the United States, paid $1.6 billion to settle charges brought by German prosecutors and the Department of Justice. It was a record fine at the time, but within Siemens' means. The company endured.

There were no mass layoffs as a result of the scandal, and for the most part innocent Siemens employees did not suffer.

Volkswagen, on the other hand, achieved a perfect trifecta of corporate malfeasance: a serious violation of law compounded by false advertising and topped off with a cover-up. Yet many of the people in charge while all that was taking place were still in charge more than a year after the fraud came to light. The supervisory board's failure to impose more serious consequences raised the cost borne by the vast majority of Volkswagen workers who had nothing to do with the scandal, as well as the communities they live in. In November 2016, Volkswagen said it would cut a net fourteen thousand jobs in Germany, about 4 percent of the domestic total, in response to dismal profit margins from sales of VW brand cars. Still, worker representatives on the supervisory board stood behind Müller. They tried to minimize the job losses, but did not use their clout to force changes in the management board.

The city of Wolfsburg was already feeling the pain by 2016. Because of Volkswagen's loss in 2015, revenue from the city's largest taxpayer by far was zero. The city was forced to raise fees for services such as public kindergartens, while postponing projects such as construction of a new fire station or improvements to the city library and a local community college. "There are 600,000 hardworking, honest people at Volkswagen," said Klaus Mohrs, the mayor of Wolfsburg. "It's terrible that the employees as well as a whole city have to suffer because a few committed fraud."

CHAPTER 22

Faster, Higher, Farther

IT IS NOT YET CLEAR what the ultimate consequences for Volkswagen will be. Perhaps car buyers will soon forget about the scandal, sales will recover, Volkswagen will contain its legal liability in Europe, and prosecutors in the United States and Europe will not find enough evidence to charge any current or former members of the management board or supervisory board. Matthias Müller will remain chief executive, giving Volkswagen time to groom a successor from within, as it has long preferred to do. The shares will recover, and so will the wealth of the Porsche and Piëch families. This is the benign scenario from Volkswagen's point of view, and the one that upper management appears to be banking on.

But it is not hard to imagine the story turning out much more harshly for Volkswagen. Say, for example, that legal campaigns in Europe pressure Volkswagen into offering owners of affected models in Europe a modest €2,000, or about $2,200, per car. That would come to €17 billion ($19 billion). Add in a few billion more for shareholder lawsuits and lawyers' fees, then combine that figure with €10 billion ($11 billion) in lost sales because of damage to Volkswagen's image. It would not be hard for the overall financial impact of the scandal to rise above $50 billion including the U.S. settlement. If Volkswagen's financial situation continues to deteriorate, it will face more downgrades by rating agencies, and its cost to borrow money will rise, further eroding profit.

At the same time, Volkswagen's ability to respond to these blows with cost cuts will be constrained by the power of worker representatives. They have already signaled they have no intention of being made to suffer for management mistakes. Despite Volkswagen's record loss in 2015, they demanded and obtained bonuses for the year of €3,950, or about $4,400. That was about €2,000 less than the year before, but still substantial under the circumstances.

Meanwhile, the crisis is eating away at the foundation of Volkswagen's success. Volkswagen diesel technology played an essential role in the company's rise to dominance in Europe. But the scandal has called attention to the human cost of diesel emissions and exposed the gulf between what cars are supposed to emit and what they really do. European regulators are already moving to make road tests a part of emissions testing in order to close the gap between theory and reality. The increased scrutiny will force carmakers to install better emissions equipment. Diesel will become more expensive and impractical for some smaller models. Volkswagen will lose one of its main competitive advantages and will have to find other ways to differentiate itself. Ironically, independent tests show that the newest model Volkswagens are among the cleanest diesels on the road. But the damage has already been done.

If its sales and profit suffer, Volkswagen will have less money to invest in new technologies like self-driving and battery-powered cars. During a time of technological shift, research and development may well determine which automakers thrive in years to come and which ones languish. Volkswagen's ability to manage these challenges will be further hampered by criminal investigations in Germany or the United States. Members of the management board or even the supervisory board must reckon with the possibility that they will be accused of complicity by former subordinates who make deals with prosecutors. Some top executives could be forced to resign, creating more turmoil.

If Volkswagen's financial state continues to deteriorate, eventually its solvency would be in doubt. Volkswagen is too important to the German economy to be allowed to fail. But European Union rules on subsidies would constrain the German government from providing aid and raise difficult questions about who should pay for a rescue. In the worst case, shareholders would be wiped out. The Porsche and the Piëch families would lose control over Volkswagen. The eighty-year bond between the company and the families would be broken.

Volkswagen has many countermeasures it can deploy before the situation becomes that dire. The company had almost €36 billion, or $40 billion, in cash and short-term investments at the end of 2015, substantially more than BMW, Daimler, or General Motors. Volkswagen could raise more money by selling assets like Lamborghini, Bentley, or the MAN and Scandia truck divisions. It could issue new shares, effectively asking shareholders to finance a rescue.

The outcome depends to a large extent on whether Volkswagen draws the right lessons from the emissions affair and makes the changes needed to prevent the scandal from becoming an existential threat. The academic world is already watching the Volkswagen case closely, because it so vividly shows how a dysfunctional corporate culture can threaten the existence of even the mightiest corporations.

The pressure to meet corporate goals at any cost is hardly unique to Volkswagen. Practically all corporate scandals stem at least in part from unrealistic targets coupled with draconian consequences for employees who fail to deliver, often combined with outsize rewards for the star performers. The risk of scandal is great if top managers do not establish clear standards of behavior and display a disregard for societal norms with their own actions.

In late 2016, the U.S. bank Wells Fargo provided yet another example of this principle in action. An investigation by the Los Angeles city attorney found that Wells Fargo sales people, under intense pressure to meet unrealistic quotas, opened accounts without customers' consent. Then they deducted fees for credit cards and other services that the customers had never ordered. When customers complained, the bank refused to provide refunds. It sent debt collection agencies after customers who didn't pay. In September 2016, John Stumpf, chief executive of Wells Fargo, appeared before a U.S. Senate committee and said he took responsibility for the unethical sales practices and apologized. But then he went on to blame the misconduct on employees, more than five thousand of whom were fired, and deny that it was the result of an orchestrated effort by the bank. "Wrongful sales practice behavior goes entirely against our values, ethics, and culture," Stumpf insisted. (He was forced to resign a few weeks later, but replaced by another insider. Stumpf walked away with about $133 million in shares, deferred compensation, and pension benefits, *Fortune* magazine reported.)

Such cases show how the right combination of fear and unrealistic expectations can turn otherwise normal employees into lawbreakers, while allowing upper managers to maintain with a straight face that they had no idea what was going on.

At Volkswagen, the rewards for people who found ways around technical problems were greater than for those who pointed out the legal risks. It is remarkable that, as far as is known, no Volkswagen employees came forward as whistle-blowers until after the fraud was exposed. Strange as it may seem to outsiders, many of those involved did not perceive that what they were doing was morally wrong. According to several former Volkswagen engineers, who spoke on condition of anonymity, many truly believed they were environmentally virtuous. They told themselves that higher nitrogen oxide emissions in diesel motors were an unavoidable trade-off in order to achieve reductions in climate-changing carbon dioxide. European Union policies reinforced this belief by actively promoting diesel.

Because Volkswagen did not do its own road testing, the engineers could delude themselves that the excess pollution was not that bad. Some were shocked by the data collected by West Virginia University and CARB showing emissions of up to forty times the legal limit. The engineers were, for the most part, the same kind of cautious middle managers one finds in many large corporations. They lived in Wolfsburg suburbs like Gifhorn in comfortable homes with compact lawns littered with children's toys. They were highly educated people with doctorates who filed patents and wrote technical papers and spoke at industry conferences. Many worked long hours because development deadlines were always extremely demanding. They were well paid but, below the level of top management, not really wealthy. Money was not what drove them. Unlike investment bankers who sold toxic mortgage-backed securities or traded on inside information, the Volkswagen employees did not, for the most part, receive big bonuses or commissions. Their goal was to keep their jobs so that they could pay their mortgages and support their families.

At least some of the people who manipulated the engine software in Volkswagens, Audis, Porsches, and other company brands knew that they were breaking the rules. But because of lack of oversight, there was no one to stop them or point out the potential consequences of their actions. Engineers are trained to solve technical problems. They are not always attuned to the collateral damage they may cause. That is what compliance departments are for. But at Volkswagen the employees responsible for making sure that engineers followed the rules lacked expertise in emissions technology. Even if the compliance people had such expertise, they lacked sufficient power inside the company to enforce their views. The company has admitted that software was not subject to review by other departments. At Volkswagen, the engineers ruled.

Volkswagen also suffered from a headquarters mentality. It paid insufficient heed to information from the field. As the company has conceded, the management board hoarded decision-making power and division heads lacked autonomy. Winterkorn was proud of his personal attention to detail. As Volkswagen expanded around the globe, Wolfsburg never loosened its grip. In the same way that it took years for Wolfsburg designers to grasp that Americans wanted larger drink holders, Wolfsburg managers failed to understand that the rules of the game were different in the United States. Use of defeat devices was a serious offense with huge potential penalties, not a misdemeanor as in Europe.

VW's myopia was not inevitable. Other German car companies recognized the need to develop a more international outlook. BMW and Daimler opened factories in the United States in the 1990s, and English was the corporate lingua franca. Volkswagen remained provincial. Top executives often had poor English skills. The only one of the nine people on the management board who was not German or Austrian was Francisco Javier Garcia Sanz, who worked in Germany most of his life. There were very few women in top management, and none on the management board until 2016.

Ultimately the responsibility for the scandal and Volkswagen's response lies with the largest shareholders, the Porsche and the Piëch families. They own a majority of the voting shares and can control the annual meeting, albeit only with the consent of the state of Lower Saxony. Ferdinand Piëch, head of one branch of the family, was a dominant presence in management for more than two decades. He created a company culture that allowed the diesel fraud to fester.

The family also has strong influence over the supervisory board. Of the ten shareholder representatives on the board, four are members of the Porsche-Piëch family. A fifth member, Hans Dieter Pötsch, the chairman, is closely associated with the family. (The other family members are Louise Kiesling, a niece of Ferdinand Piëch; Hans Michel Piëch, a brother of Ferdinand; Wolfgang Porsche, a cousin of Ferdinand and spokesman for the Porsche family; and Ferdinand Oliver Porsche, Wolfgang's brother.) Annika Falkengren, a Swedish banker, is the only shareholder representative on the Volkswagen supervisory board who is not affiliated with the families, Lower Saxony, or the sovereign wealth fund of Qatar. The other ten members of the supervisory board are worker representatives, in line with German law. There is a shortage of outside voices, and no apparent will for reform. It would be a tragedy if, eight decades after the brick walls of the Volkswagenwerk rose on the banks of the Mittelland Canal to build Ferdinand Porsche's "people's car," his descendants were the ones ultimately responsible for the Volkswagen's demise.

Epilogue

On April 26, 2016, Dan Carder put on a suit and tie and went to Lincoln Center in New York. The occasion was a party for *Time* magazine's list of the one hundred most influential people of 2016. The list included Barack Obama, Vladimir Putin, Pope Francis—and Carder. Because of his work as head of the team at West Virginia University that exposed Volkswagen, Carder achieved a degree of celebrity not usually accorded to vehicle emissions experts. But fame does not equal riches, and Carder said he still lay awake at night worrying about how to scrounge enough grants and contracts to sustain his research and testing center at West Virginia University.

Carder hoped to receive a portion of the $15 billion that Volkswagen agreed to pay as part of the settlement with U.S. diesel owners and federal and state governments. Maybe West Virginia University would get some work verifying that Volkswagen kept its promises about cutting emissions of the illegally manipulated cars. But, as of the end of 2016, Carder wasn't sure.

Judge Charles Breyer at U.S. District Court in San Francisco formally approved the $15 billion settlement on October 25, 2016. The agreement, which included about $10 billion to compensate owners, wasn't perfect. Some owners were still furious and thought Volkswagen should buy back their cars at the original purchase price. "We drank up their marketing materials and their promises of clean diesel, and paid a premium for a 2011 Audi A3 TDI," said Mark Dietrich, of San Francisco, who bought the car with his wife in 2010 just before their first child was born. "We got played the fool," Dietrich told Judge Breyer at a hearing on October 18.

But the vast majority of owners had already indicated they would accept the settlement. Breyer, while acknowledging the compromises that were made, said he wanted to put a stop to the excess pollution as soon as possible. "Cars are on the road, out of compliance with environmental regulations. And it is imperative that this matter be addressed immediately," he maintained.

The financial damage to Volkswagen from the scandal continued to worsen. In December 2016 Volkswagen agreed to a settlement applying to about 80,000 Audi, Porsche, and Volkswagen cars in the United States with three-liter motors. The settlement would cost the company $1.3 billion. Volkswagen also settled with about 105,000 owners in Canada for 2.1 billion Canadian dollars ($1.6 billion). Like their U.S. counterparts, the Canadian owners received compensation plus an option to sell their cars back to Volkswagen. Bosch, a defendant in the U.S. lawsuits, also settled, agreeing to pay $327.5 million to owners. Bosch did not admit any wrongdoing.

In October 2016, the two senior members of the often quarrelsome Porsche-Piëch clan made a show of unity. In an interview with *Der Spiegel*, Wolfgang Porsche and his cousin Hans Michel Piëch, who replaced his brother Ferdinand as the public face of the Piëch family, vowed not to intervene in daily operations of Volkswagen. The two representatives of the majority shareholders said they would take an approach different from that of Ferdinand Piëch, who was famous for meddling in daily operations (and according to his brother, still bought competing vehicles to compare them with Volkswagen products). Porsche and Hans Michel Piëch acknowledged that it was a mistake to pay top management big bonuses for 2015 and promised to reform the compensation system.

But they also signaled they did not plan a wholesale overhaul of top management. Porsche and Piëch expressed confidence in Hans Dieter

Pötsch, the supervisory board chairman who had been chief financial officer throughout the period that Volkswagen managers and engineers were deploying the fraudulent emissions software and, later, trying to cover up its existence. Pötsch had been Volkswagen's chief controller during a period when it was clear that internal controls were grossly inadequate. He was in charge of shareholder relations when Volkswagen was failing to inform shareholders about the looming risk of the emissions scandal. Porsche and Hans Michel Piëch had not lost their faith in insiders. "We have always had success by installing people from within the company into leadership positions," Piëch told *Der Spiegel*. "We have a lot of trust in Mr. Pötsch." A few weeks after the interview appeared, Volkswagen disclosed that Pötsch was being investigated by German law enforcement authorities on suspicion of concealing key information from shareholders. He remained on the job.

At Volkswagen, it was the people who came from outside who seemed to have problems. A case in point was Christine Hohmann-Dennhardt, the former German high court judge who had been brought in at the beginning of 2016 to overhaul the weak internal compliance systems that had helped breed the scandal in the first place. On January 26, 2017, the company announced that Hohmann-Dennhardt would be leaving. She and Volkswagen "are parting due to differences in their understanding of responsibilities and future operating structures within the function she leads," the company said. For a company supposedly trying to fix its corporate culture, it was not an auspicious sign.

Incredibly, evidence emerged in 2016 that some Audi vehicles sold in the United States may have contained yet another defeat device. The latest allegation involved software that managed the cars' automatic transmissions. When the software detected that the car was on rollers, it shifted differently to produce less carbon dioxide, which is also regulated. The new evidence suggested that Volkswagen and Audi had engaged in a separate case of emissions fraud and cover-up, exposing the company to yet more lawsuits and investigations. As a repeat offender, Volkswagen could expect harsh treatment from the courts and regulators. The company denied any deliberate attempt to cheat, chalking up the discrepancies to technical issues. "In the testing situation," Audi said in a statement, "dynamic shift programs can lead to incorrect readings and results that cannot be reproduced."

The court agreement resolved the federal class action civil claims against Volkswagen and Audi. But Volkswagen still faced lawsuits by twenty-one states that accused the company of violating their environmental laws. And on January 7, 2017, FBI agents served a reminder that the U.S. government was also serious about pursuing criminal charges against individual members of Volkswagen's management. Agents arrested Oliver Schmidt, the executive who had played such a prominent role in Volkswagen's dealings with CARB, at Miami International Airport. Schmidt was about to board a plane to Germany. He had decided to spend the Christmas holidays in Florida, where he owns several rental properties, despite the obvious risk that authorities would arrest him if he turned up in the United States. Schmidt apparently believed that because he had already voluntarily submitted to questioning by the FBI, they would not arrest him. Schmidt was charged with being part of a conspiracy to defraud the U.S. government and to violate the Clean Air Act. A judge ordered that Schmidt be held without bail while he awaited trial, on the grounds that he was likely to flee the country if released from custody. For the first time, the scandal had led to someone being put behind bars.

Schmidt was also the first of the defendants to give his side of the story in a United States court. In a motion filed by his lawyers in February to get him released on bail, Schmidt portrayed himself as a bit player in the conspiracy. He was not a diesel expert, according to the motion. Schmidt merely did what the experts and Volkswagen lawyers told him to do. "Mr. Schmidt's participation in these meetings was guided, at times, by internal legal advice, and given his lack of relevant technical expertise, he relied on explanations given by diesel experts," the motion said. Schmidt's argument presaged the strategy that other defendants were likely to use. They would depict themselves as cogs in the machine, at the mercy of forces more powerful than them. A judge was not convinced. On March 16 Judge Sean Cox at U.S. District Court in Detroit denied Schmidt's request for bail and set a trial date in January 2018.

A few days after Schmidt's arrest, on January 11, Volkswagen abandoned any pretense that the emissions cheating was the work of a small group of rogue engineers. After weeks of late-night negotiating sessions between government lawyers and Volkswagen's legal team, the Department of Justice announced a plea deal. Volkswagen would pay $4.3 billion in civil and criminal penalties to the government. More significantly, Volkswagen agreed to plead guilty to criminal charges. The charges included conspiracy to defraud the government and violate the Clean Air Act; obstruction of justice; and making false statements in the course of importing goods. The guilty pleas set Volkswagen apart from many banks that had been accused of wrongdoing in the wake of the financial crisis. The banks typically were able to make so-called deferred prosecution deals, in which they avoided conviction if they adhered to certain conditions. The Volkswagen deal came in the final days of Barack Obama's presidency. Finally, after practically no Wall Street bankers went to prison in the wake of the financial crisis, the Justice Department was delivering on its vow—made only days before the scandal broke in September 2015—to compel corporate wrongdoers to acknowledge their guilt.

The plea agreement included a thirty-page "Statement of Facts," in effect a detailed confession by Volkswagen of the events surrounding the emissions cheating. The document confirmed many of the allegations that had been made against the company and its executives. It described the technical impasse in 2006 that led Volkswagen managers to resort to a defeat device. It told of how higher-ups overruled engineers who objected to the illegal software. The statement explained how the defeat device was refined over the years to make it more effective. It recounted how Volkswagen executives concocted an elaborate cover-up when regulators became suspicious, and how they destroyed evidence once they realized they would be exposed. The statement of facts made it clear that a broad criminal conspiracy was at work. The people involved were high-ranking executives, compliance managers, engineers, quality control experts, software specialists, and in-house lawyers. Volkswagen was portrayed as a place where the people in charge were all too willing to resort to illegal behavior to fulfill the company's ambitions.

On January 11, the same day that Volkswagen lawyers signed the plea agreement, prosecutors in Michigan indicted Schmidt and five other Volkswagen employees, including Heinz-Jakob Neusser, the former Volkswagen development chief, and Bernd Gottweis, the Red Adair of quality control. Except for Schmidt, all the executives were in Germany, a signal that U.S. authorities would pursue the case across national borders. The documents indicated that more indictments were likely. Although no current or former members of the management board were named as suspects, Neusser and Gottweis, among others, reported directly to Winterkorn. The emissions cheating reached to the management board's doorstep.

Winterkorn, though, has continued to maintain his innocence and that of his onetime mentor, Ferdinand Piëch. Summoned on January 19 to testify before an investigative committee at the German Parliament in Berlin, Winterkorn insisted he had never heard the term "defeat device" before September 2015. He allowed that he had informed Piëch about the recall of diesels in the United States in early 2015. But Winterkorn said he hadn't told the supervisory board chairman the underlying reason for the recall because he didn't know himself. Winterkorn stuck to the position he had taken since his resignation: that he wasn't aware of any wrongdoing on his part. He professed to be mystified by suggestions that Volkswagen fostered a culture of fear which drove managers to resort to illegal measures. "Never," he said, "did I have the impression that anyone was afraid to speak an open word with me."

Soon afterward German prosecutors made it known just how skeptical they were of Winterkorn's claim. On January 28, the state attorney's office in Braunschweig expanded its investigation of the former chief executive to include alleged fraud and false advertising. After interviewing numerous witnesses and combing through documents, prosecutors concluded there was sufficient reason to believe Winterkorn may have known of the illegal software and its purpose sooner than he admitted. The prosecutors also widened their investigation of Volkswagen to include thirty-seven suspects, up from twenty-one, based in part on evidence seized during the previous months in raids on dozens of homes and offices in the Wolfsburg area and elsewhere. According to German press reports, one of the homes raided was Winterkorn's villa in Munich. Winterkorn continued to maintain his innocence, issuing a statement saying he stood by his testimony in Berlin.

Piëch had managed to remain practically invisible since Volkswagen's scandal became public. But in February 2017 the man who had for so long dominated the company once again became the focal point of attention and controversy. *Der Spiegel* magazine reported that Piëch had said during questioning by German investigators in December 2016 that he learned of a major emissions problem in the United States in February 2015, while still chairman of the Volkswagen supervisory board. What's more, according to *Spiegel*, Piëch had told investigators that he had later passed on the information to the supervisory board's executive committee, which included Stephan Weil, the prime minister of Lower Saxony, and Wolfgang Porsche. Winterkorn had assured Piëch that there was nothing to worry about, according to the *Spiegel* report.

If true, Piëch's statement was explosive. Volkswagen's legal liability would be much greater if it turned out that people at the very top of the company knew of the emissions wrongdoing early in 2015 and failed to take appropriate action. Volkswagen did not dispute *Spiegel*'s account of what Piëch had told German prosecutors. But, in a statement, the Volkswagen supervisory board said it "emphatically repudiates" Piëch's assertion that he had informed some members about the emissions issue. Internal investigators from the Jones Day law firm had examined the accusations, Volkswagen stated, and "no evidence was forthcoming indicating the accuracy of these allegations, which were classified as implausible overall." Winterkorn would not comment until he had a chance to examine prosecutors' files, his lawyer, Felix Dörr, said.

Piëch's reasons for implicating other members of the supervisory board were obscure. By admitting that he had been informed of the emissions problem in early 2015, Piëch was potentially exposing himself. He was not obligated to say anything—during interviews with prosecutors he could have exercised his right not to incriminate himself. Piëch was also endangering his own wealth. By raising the financial risk to Volkswagen, Piëch potentially undercut the value of his substantial stake in the company. Not for the first time, his motive was unclear. Perhaps the explanation was as simple as the one put forward by a writer for the *Frankfurter Allgemeine Zeitung*: revenge.

Piëch caused further turmoil when it emerged on March 17 that he was trying to sell his stake in Porsche Automobil Holding, the vessel for the family's Volkswagen shares. Piëch was obligated to offer the roughly 15 percent stake, worth €1.1 billion (or about $1.2 billion), to other family members. But if they couldn't come up with the money, Piëch could sell his shares to an outsider (for example, a Chinese investor seeking access to the European auto market). Piëch's move threatened to weaken the Porsche family's power over Volkswagen and create further uncertainty. But it also raised the prospect that the company would gain some badly needed outside oversight.

Volkswagen continued to maintain that no one at the top level of operating management was culpable in the fraud. At a news conference in Wolfsburg with other members of the management board on March 14, Matthias Müller said that the admissions Volkswagen made as part of the U.S. plea agreement did not involve "anyone sitting at this podium." That group included Rupert Stadler, the chief executive of Audi and a member of the Volkswagen board.

But even as Müller spoke to reporters, prosecutors and police investigators in Munich were secretly preparing a major operation centered in Ingolstadt, Audi's home base. While the state's attorney in Braunschweig had been taking the lead in investigating Volkswagen, a team in Munich had been quietly probing Audi's involvement in the fraud in the United States. The day after Müller's press conference the Munich officials struck, appearing at Audi headquarters at 7 a.m. with a search warrant. The timing was exquisitely bad from Audi's point of view—it had scheduled its own press conference that same morning to talk about its 2016 financial results. Stadler and other members of the Audi management board had to answer awkward questions from the assembled journalists about the searches taking place in the Audi offices nearby.

From Volkswagen's point of view, the news got even worse. The Munich authorities also searched offices belonging to Jones Day, the law firm conducting Volkswagen's internal investigation. The action suggested that the German investigators doubted whether Jones Day had turned over all the relevant information it had collected. It was very unusual, but not unheard of, in Germany for investigators to raid the offices of a law firm working for a client under investigation. German laws on attorney-client privilege are not as strict as those in the United States. Ansgar Rempp, the partner in charge of Jones Day in Germany, declined to comment.

What's more, the search warrant signed by a Munich judge authorized searches of offices belonging to Müller and Stadler. Investigators had authority to seize correspondence, appointment calendars, notebooks, e-mails, mobile phones, and even passwords from Müller, who had spent most of his career at Audi and served as chairman of the division's supervisory board, as well as from Stadler, who had overseen Audi for a decade. The search warrant made no allegations against the two executives or any other individuals. Investigators were still trying to figure out who devised the illegal software in Audis sold in the United States, the warrant said, as well as "which levels of company hierarchy were informed, and what level made the decision to mass produce the defeat device." Still, the searches were an indication of the degree to which the scandal continued to roil Volkswagen a year and a half after the EPA had first issued its notice of violation against the company.

Volkswagen did not give up on the United States. On the contrary, the company doubled down on America, going ahead with plans to build a new seven-seat SUV in Chattanooga called the Atlas. Volkswagen invested $900 million in the Tennessee factory to accommodate production of the new vehicle, which was scheduled to go on sale in May 2017. The Atlas was available with a 2-liter, four-cylinder motor or a 3.6 liter, six-cylinder engine, both powered by gasoline. There was no diesel option.

Afterword

If you were a German concerned about your country's future, the worst-case outcome of the Volkswagen scandal would be that the taint would spread to other carmakers and threaten one of the pillars of the economy, perhaps even undercutting Germany's reputation for engineering excellence. Two years after the scandal came to light, that is exactly what had happened.

Gradually, it dawned on the European public how fundamentally the automakers had deluded them about the risks to human health from diesel and the chasm between emissions as advertised and emissions in practice. The German carmakers, including BMW and Daimler's Mercedes-Benz division, bore the brunt of consumer wrath. It didn't help their reputations when the European Commission's antitrust unit said in July 2017 that it was investigating whether Daimler, BMW, Volkswagen, Audi, and Porsche colluded illegally to hold down the prices of some technology. The investigation included searches of the carmakers' offices. The magazine *Der Spiegel* reported that the collusion might have included emissions technology, raising the possibility that the carmakers had agreed to limits on the effectiveness of anti-pollution equipment, laying the groundwork for the scandal. All the carmakers said they were cooperating with investigators; BMW went further and issued a statement denying that it had ever cheated on emissions.

Battered by the negative publicity, diesel suffered a precipitous decline in market share in Europe, where it had once been the most popular engine option. The implications for the German economy were ominous. The consumer backlash undercut the German carmakers' business model and, by extension, their dominance of the luxury car market. Auto manufacturers and suppliers account for about 2 out of 100 jobs in Germany, and some of these jobs would be in danger if buyers continued to reject diesel. "The so-called diesel crisis must be regarded as a new risk for the German economy," the German Finance Ministry warned in August 2017.

The scandal quickly took on political dimensions. Initially, many Germans had seen the accusations against Volkswagen as a plot by the U.S. government to undercut foreign competitors. But in 2017 public opinion pivoted. It became increasingly obvious that German elected officials across the political spectrum had enabled excess emissions by blocking efforts to tighten enforcement and turning a blind eye as carmakers made a mockery of emissions standards. The appearance of impropriety was not helped by the fact that there was a revolving door between government and the car industry. For example, Thomas Steg, Volkswagen's top liaison with government, was a former spokesman for Angela Merkel, the chancellor; Eckart von Klaeden, responsible for government relations worldwide at Daimler, had previously served under the chancellor as a junior minister.

Many German politicians had functioned as de facto lobbyists for the country's most important industry. That included the chancellor herself. Mary Nichols, the chairwoman of the California Air Resources Board, told an illuminating anecdote about Merkel while testifying in March 2017 via video link to an investigative committee of the German Parliament. During a visit to Sacramento in 2010, Nichols testified, Merkel met with Arnold Schwarzenegger, then the governor of California, to talk about climate change. Nichols joined them. No sooner had the meeting begun, Nichols testified, when Merkel began complaining about the state's strict limits on nitrogen oxides. "She said, 'Your nitrogen oxide limits are too strict, and that is hurting our

German diesels,' " Nichols said. "She was there, it seemed, as spokeswoman for the auto industry." Nichols recalled that she had told the chancellor diplomatically that, while California standards were indeed tough, they had to be tough to ensure clean air, and that she had confidence that the German carmakers would be able to meet them. Merkel then changed the subject, Nichols said. "For me it was very surprising that she knew anything at all about the German manufacturers' NOx problems," Nichols said. "Never before or since have I experienced a similar intervention by a politician against our environmental laws."

The chancellor didn't deny bringing up nitrogen oxides when she met with Schwarzenegger and Nichols, but she told the story differently. When Merkel testified before the Bundestag committee, she pointed out that she had previously been the German minister of the environment, and thus was well acquainted with emissions technology. Merkel had argued in favor of diesel as a way of reducing carbon dioxide emissions and combating climate change, she said. "German diesels didn't have a chance in California," Merkel recalled saying. "Is that really a good strategy to protect the climate?"

Advocacy for the German auto industry once would have been seen as patriotic. But in the aftermath of the emissions scandal it became a serious political liability for Merkel, who was in the midst of a tough election campaign. She tried to distance herself from the automakers. "I'm just as angry about the fraud as you," she told *Der Spiegel* in September 2017, weeks before national elections. Her government summoned the chief executives of Volkswagen, BMW, and Daimler—Matthias Müller, Harald Krüger, and Dieter Zetsche—to Berlin for an emergency summit. The auto bosses announced incentives for owners of older diesels to trade them in for new models, and they said they would contribute to a fund worth 500 million euros, or $590 million, to reduce urban pollution by modernizing bus fleets, building bike paths, or other measures. But they rejected proposals to upgrade emissions systems in cars already on the road or to compensate people for the declining value of their diesels.

As political damage control, the meeting was a failure. Both Merkel's

Christian Democratic Union and the Social Democrats, her partners in the parliamentary coalition that ran Germany, suffered big declines in support when voters went to the polls a few weeks later. Support for the Christian Democrats slumped to 33 percent, compared to 42 percent in the 2013 elections. The Social Democrats won 21 percent compared to 26 percent in 2013. Merkel's party still got the most votes and she remained chancellor, but she governed with a much weaker mandate. The diesel scandal was by no means the only reason for the disastrous result—Merkel's open-door policy toward immigrants pushed some conservative voters to a new far-right party. But diesel contributed to the feeling among many voters that their leaders had let them down, and it helped redraw the political map.

As public disaffection built, numerous European cities began discussing bans on diesel cars. That got consumers' attention; no one wanted to buy a car that would have to be parked at the city limits. In October 2017, London announced it would impose an extra congestion charge of £10, or $13.20, per day for older diesels—primarily those built before 2005—to drive in the city center. That was on top of the existing congestion charge of £11.50 per day, bringing the total cost of driving a diesel clunker into London to £21.50, or almost $30. The mayor of London, Sadiq Khan, blamed bad air, especially diesel exhaust, for the asthma that he developed as an adult. In an illustration of the depth of public outrage, some of the cities contemplating similar restrictions were auto industry bastions. They included Munich, home of BMW, and Stuttgart, home of Daimler and Porsche. By late 2017 the market share of diesel, an essential component of the German carmakers' strategy, was in free fall. Automakers protested that the latest generation of diesels really was clean. But the car companies had squandered so much trust it was unlikely diesel's reputation would ever recover.

The decline of diesel was particularly a threat to the makers of big luxury cars and SUVs—Audi, Porsche, BMW, and Daimler. Big cars were very difficult to sell in Europe without the fuel savings provided by diesel. In response, carmakers began a crash program to electrify their fleets. Volvo Cars, another maker of big luxury cars that depended heavily on diesel, announced that it would not develop any new internal combustion-power vehicles after 2019. A visitor to the International Motor Show in Frankfurt in September 2017—the first time the show had been held in Germany since the Volkswagen scandal had cast a shadow over the industry two years earlier—might have gotten the impression that European roads were already teeming with hybrids and battery-powered cars. In fact, the Germans were frantically trying to make up time lost to Toyota with its hybrids and Tesla with its electric cars. In Kamenz, a sleepy town in eastern Germany, Daimler began building a massive 500 million euro ($600 million) factory solely to produce batteries for hybrids and electric cars. Not far away, on the other side of the border with the Czech Republic, prospectors drilled for ore containing the lithium needed for battery cells. It was a modern gold rush, driven by carmakers' desperation to avoid being overrun by technological change. If there was a silver lining to the diesel scandal, it was that it inspired the German carmakers to wholeheartedly devote their formidable manufacturing networks and research and development prowess to emission-free vehicles.

Criminal investigations in Germany and the United States moved forward slowly. In August 2017, James Liang, the engineer based in California who had been the first to plead guilty to charges stemming from the diesel fraud, was sentenced to 40 months in prison, followed by two years of supervised release and a $200,000 fine. Judge Sean F. Cox at U.S. District Court in Detroit imposed a harsher penalty than prosecutors advised, saying that Liang had been part of a "massive and stunning" fraud. Liang, sixty-three, became inmate number 54762-039 at Taft Correctional Institution, a medium- and minimum-security facility about 115 miles north of Los Angeles. (Jordan Belfort, the stock market swindler of *The Wolf of Wall Street* fame, also did his time at Taft.) Liang was scheduled to be released on August 28, 2020.

An even harsher fate awaited Oliver Schmidt, the former point man for Volkswagen in its ultimately futile attempt to mislead the California Air Resources Board and the Environmental Protection Agency. Schmidt, who had been arrested in a men's restroom at Miami

International Airport in January 2017, pleaded guilty the following August to charges that included conspiracy to defraud the United States government and conspiracy to violate the Clean Air Act. In his plea agreement, Schmidt admitted that other, unnamed Volkswagen employees had informed him about the defeat devices in summer 2015, and that he had taken part in discussions about how to answer questions from regulators without revealing the fraud. Schmidt also admitted feeding false information to a California Air Resources Board employee—Alberto Ayala—during the meeting in Traverse City, Michigan, described in Chapter 17. Schmidt received a sentence of seven years in prison and a $400,000 fine when he faced Judge Cox on December 6. When Schmidt was done serving his sentence, he would be deported.

Most of the suspects, however, were in Germany, out of reach of the Justice Department. Ultimately, it would be up to the German prosecutors to determine which engineers and executives took part in the conspiracy and which ones were guilty of crimes. With more than 50 suspects, it was a complex task. By the end of 2017, more than two years after the Environmental Protection Agency issued the initial notice of violation against Volkswagen, prosecutors in Braunschweig had still not made any arrests or filed any formal charges, and did not appear to be close to doing so. Under German procedures, before states' attorneys filed charges, lawyers for suspects would have a chance to examine the evidence against their clients. That had not happened by late 2017, an indication that any charges were still months away, if not longer.

Prosecutors in Munich proved to be more aggressive than those in Braunschweig, Volkswagen's home turf. The Munich state's attorney was responsible for investigating Audi executives' role in the diesel fraud. Ingolstadt, Audi's home base, lay within their geographical jurisdiction. In July 2017, the Munich prosecutors arrested Zaccheo Giovanni Pamio, who had been head of thermodynamics in Audi's engine development department. Though Pamio had spent most of his career in Germany, he was a citizen of Italy and therefore did not enjoy the immunity from extradition that Germany grants its own citizens.

Pamio was held without bail in a jail near Munich while U.S. authorities initiated extradition proceedings. He was the first person to be arrested in Germany in connection with the diesel fraud, and under the pressure of confinement he quickly displayed a willingness to cooperate. During at least ten lengthy interrogation sessions, Pamio told the Munich prosecutors that top Audi managers were aware as early as 2006 that vehicles made by the division could not meet American and European emissions standards. As described in Chapter 10, the cars could not carry an adequate supply of the AdBlue urea solution used to neutralize nitrogen oxide emissions. Rather than install a bigger AdBlue holding tank, or inconvenience owners with more frequent refills, Audi installed a defeat device. Details about what Pamio was telling prosecutors came from Walter Lechner, the veteran Munich defense lawyer who represented him. In Germany, or the United States for that matter, it is very unusual for lawyers to discuss an investigation in progress. Lechner's decision to talk to the author and other journalists was a clear signal that he was not going to allow Pamio to serve as Audi's scapegoat. Pamio gave investigators emails and other documents to corroborate what he was telling them, Lechner said. Pamio also spoke to U.S. investigators. Spokesmen for Volkswagen and Audi declined to comment on Pamio's allegations.

The information that Pamio provided helped the investigators build a case against people higher up in the organization chart. In late September 2017, Munich prosecutors arrested Wolfgang Hatz, the former head of engine development at Volkswagen who had worked closely both with Martin Winterkorn and with his successor as chief executive, Matthias Müller. Hatz, who had been suspended soon after the scandal came to light, was the highest-ranking executive to be arrested up to that point, and his incarceration brought the scandal to the management board's doorstep. (Hatz was the same person, described in Chapter 10, who had been caught on camera in 2007 griping that California's air quality standards were "too aggressive.") Hatz was also held without bail, which is possible in Germany if a judge determines that the suspect would flee or might use his freedom to obstruct justice.

Prosecutors were certain to put pressure on Hatz to testify against others.

Volkswagen continued to insist that the fraud was the work of people below the management board and that no one in the top layer was involved. Müller showed no inclination to resign nor did Rupert Stadler, the chief executive of Audi. Substantiating allegations against people at the top of any big corporation is difficult because they often do not write emails or put their names on documents. Assistants do that for them. There is no paper trail. But the more information that came out about the fraud the clearer it became that dozens if not hundreds of people took part, including many people who reported directly to the management board. After the hardcover edition of this book went to press, the author was able to view documents that circulated in the months after the meeting in Volkswagen's Research and Development building in November 2006, described in Chapter 10, at which executives decided to create the emissions cheating software. Internal memos and presentations in the months that followed the meeting showed that executives just below Winterkorn and Müller openly discussed the illegal technology. They also wanted top management's approval before deploying it.

For example, in May 2007, one status report on the progress of the EA 189, the new diesel engine designed with the United States market in mind, noted that high-echelon approval was needed for "functionality off-cycle." That terminology could be interpreted as code for software that would allow cars to produce more emissions outside the simulated driving cycle used by regulators to test cars in a lab. The requests for management board approval later disappeared from the status reports, but the documents show that engineers continued to struggle with the emissions issue. By November 2007, the company's plans to regain market share in the United States with "clean diesel" were seriously behind schedule and over budget. Dealers were impatient to begin selling the new cars. Winterkorn scheduled an emergency meeting to discuss the problem.

Ahead of the meeting, executives reporting to Winterkorn prepared a PowerPoint presentation that described a software "package" that would optimize emissions "in the cycle." It was reasonable to conclude that the Volkswagen executives were describing software that would behave differently under test conditions, a defeat device. One page of the presentation described how, for example, the software would detect the procedure used by testers to prepare the engine for testing, and detect when they adjusted the rollers in the lab. The software would also ensure that the car's onboard diagnostics system would not report the measures.

On November 7, 2007, a day before the planned meeting, the presentation was sent as an email attachment to more than a dozen managers just below the top layer of management. The email included a warning to recipients not to forward it to anyone else. That evening, a second presentation was distributed to a smaller group, including several managers, although not Winterkorn or Müller. Both presentations noted that the new engine could not meet nitrogen oxide standards with the existing equipment and proposed improvements that would cost about 270 euros, or about $300, per car. But the second presentation did not include the appendix describing the specifics of the defeat device. Volkswagen maintained that Müller never saw the earlier version of the presentation that detailed the illegal software. "The Volkswagen Group is aware of the documents," the company said in an email to the author, "and they do not support the inference that Matthias Müller knew about efforts to develop and use the defeat device." As previously noted, Winterkorn has denied wrongdoing. An internal summary of the meeting, reviewed by the author, gave only a vague description of the discussion. "The situation regarding technology was acknowledged, and the implementation of the presented measures was confirmed," the meeting minutes said. After the meeting, Volkswagen delayed sale of the new motor in the United States by a year, to late 2008. The company made some improvements to the motors' emissions systems, but as subsequent events showed, they were not enough.

Even taking Volkswagen's denial at face value, the documents indicate that a large number of people in close proximity to Müller and

Winterkorn were aware of the illegal software in 2007 and discussed it intensively. To accept Volkswagen's version of events, one has to believe that these executives managed to avoid mentioning the software cheat during a meeting whose purpose was to find a solution for the emissions problem. Further, they would have had to convince Winterkorn and Müller, who at the time was Volkswagen's head of product planning, that they had found some other, legal solution to this extremely difficult technical quandary. Given the two men's knowledge and attention to detail, it is hard to imagine that they would be so easily deceived. If the lower-ranking engineers covered up the defeat device, they would have had to keep this explosive secret from top management for the next eight years, even as the illegal software was installed in millions of vehicles.

In the author's opinion, that scenario is not credible. But determining beyond a reasonable doubt what really happened is a daunting task for investigators. Whether they will have the resources, the will, and above all the evidence to pursue top managers remains, at this writing, an open question.

There will be serious financial consequences for Volkswagen if it can be proved that any members of the management board were complicit. Shareholders in Europe and the United States who were suing the company, seeking more than $10 billion in damages, would have more ammunition for their claim that the company was derelict in its duty to inform them about the risks that Volkswagen faced from its emissions technology. As it was, the financial burden of the fraud continued to rise. Volkswagen subtracted 2.6 billion euros, or $3 billion, from its earnings in the third quarter of 2017 to reflect the unexpectedly high cost of fixing diesels in the United States to comply with the terms of the class-action settlement. That brought the total cost just in the United States to $26 billion, not counting lawyers' fees. The number of lawsuits in Europe continued to grow, despite the obstacles to class-action suits. In Germany the number of lawsuits by individual diesel owners grew to 4,600 by late 2017, and there were class-action suits under way in Britain and Australia. In contrast to the United

States, Volkswagen refused to settle in Europe. The company argued that what it had done was not illegal in Europe, as implausible as that argument seemed. The strategy appeared to be to draw the cases out as long as possible, in order to postpone the day of reckoning. Volkswagen seemed to be hoping that, as time passed, the world would gradually lose interest.

There were signs the strategy would succeed. Stock market investors, at least, believed the company was overcoming the scandal. By late 2017 Volkswagen's preferred shares, the ones most actively traded, had come close to making up the ground they had lost since September 18, 2015, when the world learned of the emissions cheating. Sales in Russia and Latin American markets began to recover as economic growth in those regions improved. Even in the United States, where the damage to Volkswagen's reputation was most severe, sales rose in 2017 in part because of the popularity of the Atlas SUV made in Chattanooga. As a result, Volkswagen could report a slight increase in global vehicle sales for the year. For investors, those signs of Volkswagen's resiliency outweighed some disquieting developments, such as slippage in market share in the European Union and the lack of any convincing reform of company governance and culture.

The revival of Volkswagen shares may also have reflected the company's commitment to electric vehicles, which was genuine. The emissions scandal forced Volkswagen to abandon its longtime obsession with diesel. Instead, Volkswagen focused its still formidable industrial might on a transition to electric cars. That was the only way the company could rebuild its reputation and comply with increasingly stringent limits on carbon dioxide emissions. Volkswagen promised to introduce a battery-powered car in 2019 that would be roughly equivalent to the Golf in size and affordability. The new vehicle would be just the first in a series of electrified models, including a van and even a battery-powered Porsche. The models would be designed from the ground up as electric vehicles, not just as adaptations of existing models. That allowed engineers to take advantage of the freedom that comes from no longer having to configure the car around a bulky internal combustion engine and transmission. Volkswagen engineers designed a so-called modular electric drive kit, a collection of components that could be deployed in all of the company's brands. It was the same platform strategy implemented by Ferdinand Piëch decades earlier, now in the service of emission-free vehicles.

Volkswagen's electric vehicle push sounded like a public relations strategy when executives first started talking about it in 2015. But two years later there was no denying that Volkswagen was committing real money. In November 2017 the company announced that it would invest 34 billion euros, or $40 billion, through 2022 in battery-powered vehicle technology and self-driving cars. Some of the money would be used to convert a factory in Zwickau, in eastern Germany, to produce nothing but battery-powered vehicles. Volkswagen's announcement came at the same time that Tesla was suffering from problems rolling out its Model 3, the California carmaker's most affordable model to date. More than half a million people had paid $1000 apiece to reserve Model 3s—a sign that there was keen demand for a well-designed electric car—but Tesla had managed to manufacture only a few hundred. Another explanation for the increase in Volkswagen's share price was that investors began to believe that it could do what Tesla couldn't: quickly ramp up production of affordable, appealing electric vehicles.

Few companies anywhere do bigness better than Volkswagen. The same empire building and grandiose ambition that had gotten Volkswagen in so much trouble in the first place could, if used to populate the world's roads with emission-free vehicles, turn out to be good for mankind.

ACKNOWLEDGMENTS

This book would not exist if not for the initiative of John Glusman, editor in chief of W. W. Norton, who approached me about writing it only weeks after the Volkswagen story broke, somehow divining what a good story it would turn out to be. I am grateful not only for the opportunity that John gave me but also for the wisdom and encouragement he provided during the research and writing. Marly Rusoff, my agent, handled the business side of the project with aplomb and efficiency, informed by her genuine love of the printed word.

I am indebted to the many people who agreed to tell me their side of the Volkswagen story. All the on-the-record sources are named in the text or endnotes, and I would like to thank each one. There are some sources who could not be named. To them I owe a special debt for the risks they took to their careers and reputations by speaking to me. Manfred Grieger at the Volkswagen archives helped guide me to useful documents, and his book with Hans Mommsen on Volkswagen during the Nazi period was a crucial resource.

My editors at the *New York Times* allowed me to make Volkswagen the focus of my work for a year. The *Times*'s unparalleled news infrastructure underpinned my reporting. I would like to thank the editors who have overseen coverage of Volkswagen, in particular Dean

Murphy, Adrienne Carter, Jesse Pesta, Tim Race, and Prashant Rao. I was by no means the only reporter covering the Volkswagen story for the *Times*. Many others made crucial reporting breakthroughs or were essential collaborators and collegial co-workers. They include Danny Hakim, Hiroko Tabuchi, Melissa Eddy, Alison Smale, Graham Bowley, Bill Vlasic, and Coral Davenport. Danny also provided much valuable advice. I have tried to be scrupulous about properly noting when I drew on others' published work, but if I have inadvertently failed to give proper credit, please forgive me.

My wife, Bettina Stark, was a steady source of love and emotional support. She applied her fine sense of aesthetics to the selection of the photos found in this book. She fended off distractions so that I could concentrate on writing. Saskia, our daughter, is a constant joy and inspiration. I could not have finished this book without them.

And finally, thanks to my parents, Zita Lee and John Ewing, for the example they provide to this day in optimism and perseverance.

NOTES

CHAPTER 1: ROAD TRIP

4 **Carl Benz, who filed a patent**: "Company History: Benz Patent Motor Car, the first automobile (1885–1886)," Daimler website, https://www.daimler.com/company/tradition/company-history/1885-1886.html.

CHAPTER 2: THE GRANDSON

5 **At least since the 1920s**: Porsche Museum, *Ferdinand Porsche and the Volkswagen* (Stuttgart: Porsche AG, 2009), 13.

5 **In 1938, there was only**: Otto D. Tolischus, "Nazi Hopes Ride the 'Volksauto,'" *New York Times*, Oct. 16, 1938, p. 129.

5 **Hitler seized on the idea**: Porsche Museum, *Porsche and the Volkswagen*, 15.

5 **The car, he decreed**: Ibid., 19.

5 **The Führer wanted**: Ibid., 17–19.

6 **A native of the Sudetenland**: Ibid., 12–19.

6 **Hitler invited Porsche**: Ibid., 18.

7 **The Type 32 had a body design**: Ibid., 24.

7 **In Munich, the group met Hitler**: Ibid., 41.

7 **The rivals insisted**: Ibid., 47.

7 **The hostility of the established manufacturers**: Ibid., 62.

7 **After the war, in return**: Berthold Huber, "Die historische Verantwortung für VW," *Frankfurter Allgemeine Zeitung*, March 5, 2008 (all translations by the author unless otherwise noted).

8 **monumental temper tantrums**: Ferdinand Piëch, with Herbert Völker, *Auto.Biographie* (Hamburg: Hoffmann und Campe Verlag, 2002), 20.

8 **Porsche had visited America**: Ibid., 23.

8 **The site chosen for the new plant**: Porsche Museum, *Porsche and the Volkswagen*, 91.

8 **Hitler himself spoke at the ceremony**: Ibid.

9 **a lawsuit against Porsche was settled**: Steven Parissien, *The Life of the Automobile: The Complete History of the Automobile* (New York: St. Martin's Press, 2014), 122.

9 **a *New York Times* correspondent**: Porsche Museum, *Porsche and the Volkswagen*, 91.

9 **described the car as a "beetle"**: Ibid., 96.

9 **the biggest factory of any kind**: Tolischus, "Nazi Hopes Ride."

10 **The 336,000 Germans**: Markus Lupa, *Volkswagen Chronik: Der Weg zum Global Player* (Wolfsburg: Volkswagen AG, 2008), 8.

10 **Instead of making Volkswagens**: Ibid., 10.

10 **The factory also produced**: Peter von Bölke, "Der Führer und sein Tüftler," *Der Spiegel*, Nov. 4, 1996, accessed Nov. 22, 2015, http://www.spiegel.de/spiegel/print/d-9114600.html.

10 **There was even an amphibious**: Porsche Museum, *Porsche and the Volkswagen*, 158.

10 **The tank was a famous failure**: William Manchester, *The Arms of Krupp, 1587–1968* (Boston: Little, Brown, 1968), 441.

11 **Years later, a surviving example**: Piëch, *Auto.Biographie*, 81.

11 **In the summer of 1942**: Ibid., 24.

11 **Ferdinand Piëch later recalled causing**: Ibid., 9.

12 **The KdF Wagen factory was not bombed**: Ibid., 23.

12 **the factory made extensive use**: Lupa, *Volkswagen Chronik*, 10.

12 **The Porsches, who lived most**: Piëch, *Auto.Biographie*, 10.

13 **"I would like to someday work"**: Ibid., 11

13 **Piëch later excused his grandfather's role**: Ibid., 24

13 **Even the best-treated among them**: Bölke, "Der Führer und sein Tüftler."

14 **"eventually they all died in Rühen"**: Sara Frenkel, in *Überleben in Angst: Vier Juden berichten über ihre Zeit im Volkswagenwerk in den Jahren 1943 bis 1945* (Wolfsburg: Volkswagen AG, Historische Kommunikation, 2005), 66.

14 **Porsche had "sleepwalked"**: Bölke, "Der Führer und sein Tüftler."

14 **Volkswagen used an unusually large**: Hans Mommsen, "Zwangsarbeit im Dritten Reich: Eine Einleitung," in *Erinnerungsstätte an die Zwangsarbeit auf dem Gelände des Volkswagenwerks* (Memorial for forced laborers on the grounds of the Volkswagen factory) (Wolfsburg: Volkswagen AG, 2014), 3.

14 **Porsche and Piëch family members also complained**: Piëch, *Auto.Biographie*, 25.

14 **The first major Allied attack**: Eighth Air Force Historical Society, "WWII 8th AAF Combat Chronology," http://www.8thafhs.org/combat1944a.htm.

14 **Bombers of the U.S. Eighth Air Force**: Hans Mommsen and Manfred Grieger, *Das Volkswagenwerk und seine Arbeiter im Dritten Reich* (Düsseldorf: ECON Verlag, 1997), 927–28, 634.

15 **Much of the machinery**: *Erinnerungsstätte an die Zwangsarbeit*, 122.

15 **The Volkswagenwerk continued**: Hans Mommsen, "Das Volkswagenwerk und die 'Stunde Null': Kontinuität und Diskontinuität," website for the exhibition "Aufbau West, Aufbau Ost" (Berlin: Deutsches Historische Museum, 1997), accessed Dec. 6, 2015, http://www.dhm.de/archiv/ausstellungen/aufbau_west_ost/katlg14.htm.

15 **"The Americans!"** *STO à KdF: Die Erinnerungen des Jean Baudet, 1943–1945* (Wolfsburg: Volkswagen AG, Unternehmensarchiv, 2000).

15 **Ferdinand Porsche was long gone**: Mommsen, "Zwangsarbeit im Dritten Reich."

15 **Piëch took a substantial amount**: Mommsen and Grieger, *Volkswagenwerk*, 927.

16 **Piëch and the Porsches even dismantled**: Letter from Porsche Konstruktion to Volkswagen regarding billing disputes, June 18, 1948, Volkswagen Archive.

16 **the fate that befell Opel**: Opel website, http://www.opel.de/opel-erleben/ueber-opel/tradition.html.

17 **"The responsible officers"**: Markus Lupa, *Spurwechsel auf britischen Befehl: Der Wandel des Volkswagenwerks zum Marktunternehmen, 1945–1949* (Wolfsburg: Volkswagen AG, 2010), 23.

17 **Like other German industrialists**: Parissien, *Life of the Automobile*, 183–84.

17 **After being questioned by Allied interrogators**: Mommsen and Grieger, *Volkswagenwerk*, 940–44.

18 **Porsche benefited from testimony**: Mommsen and Grieger, *Volkswagenwerk*, 943–44.

CHAPTER 3: RENAISSANCE

20 **a resourceful and energetic major**: Ralf Richter, *Ivan Hirst: Britischer Offizier und Manager des Volkswagenaufbaus* (Wolfsburg: Volkswagen AG, 2003), 15–17.

20 **five cars delivered to a dealer**: Ibid., 84.

20 **He pushed to improve the quality**: Markus Lupa, *Volkswagen Chronik: Der Weg zum Global Player* (Wolfsburg: Volkswagen AG, 2008), 17.

20 **Hirst also created a twelve-person**: Markus Lupa, *Changing Lanes under British Command: The Transformation of Volkswagen from a Factory into a Commercial Enterprise, 1945–1949*, trans. Patricia C. Sutcliffe (Wolfsburg: Volkswagen AG, 2011), 62.

21 **Nordhoff insisted later**: Richter, *Ivan Hirst*, 111.

21 **production of Volkswagens**: Lupa, *Volkswagen Chronik*, 15.

21 **Nordhoff brushed him off**: Richter, *Ivan Hirst*, 109.

21 **In the years before his death**: Steven Parissien, *The Life of the Automobile: The Complete History of the Automobile* (New York: St. Martin's Press, 2014), 183.

22 **Nordhoff proposed that Volkswagen**: Protokoll Hauptabteilungsleiter Besprechung, Oct. 18, 1951, Volkswagen Archives, Wolfsburg.

22 **"so that unfriendly articles"**: Ibid., Aug. 31, 1951.

23 **Nordhoff set the precedent**: Hans Mommsen and Manfred Grieger, *Das Volkswagenwerk und seine Arbeiter im Dritten Reich* (Düsseldorf: ECON Verlag, 1997), 938–39.

23 **British occupation forces objected**: Ibid., 939.

23 **They also noted that a Jewish prewar shareholder**: Memorandum British Property Control Branch to Volkswagen, July 25, 1949, Volkswagen Archives.

23 **But when the issue of patents**: Letter from Volkswagenwerk managers to Stuttgart Finanzgericht (Finance Court), June 9, 1948, Volkswagen Archives.

24 **"The whole sphere in which we work"**: Letter from a Herr Knott to Ferry Porsche, 1949 (precise date missing), Volkswagen Archives.

24 **Left to pursue his claim**: Eberhard Reuß, "Adolf Rosenberger: Porsches dritter Mann und ein wenig ruhmreiches Kapitel der Firmengeschichte," transcript of documentary broadcast on SWR2, March 12, 2010, http://www.swr.de/swr2/programm/sendungen/tandem/-/id=10068636/property=download/nid=8986864/tyumep/swr2-tandem-20120906-1005.pdf.

25 **Nordhoff should come to Stuttgart**: Letter from Ferry Porsche to Heinrich Nordhoff Oct. 2, 1953, Volkswagen Archives.

CHAPTER 4: THE SCION

27 **Ferdinand attributed his father's**: Ferdinand Piëch, with Herbert Völker, *Auto.Biographie* (Hamburg: Hoffmann und Campe Verlag, 2002), 34–37.

27 **In 1959, Piëch enrolled**: Ibid., 40,

28 **He married his pregnant girlfriend**: Ibid., 41–43.

28 **Piëch hoped to work with airplanes**: Ibid., 43.

28 **They had better things to do**: Ibid., 42.

28 **he was comfortable enough**: Ibid., 43.

29 **There was just one condition**: Ibid., 56.

29 **Three died in cars by other**: Ibid., 59.

30 **Within the Porsch-Piëch clan**: Ibid., 66.

30 **By 1970, four . . . grandchildren**: Ibid., 67.

30 **Porsche was by then**: Anton Hunger and Dieter Landenberger, *Porsche Chronicle, 1931–2004* (Munich: Piper Verlag, 2008), 78.

30 **Though its products were world renowned**: Ibid., 75.

30 **In the fall of 1970**: Piëch, *Auto.Biographie*, 69–71.

31 **Certainly, he did nothing to promote**: Ibid., 80.

31 **Now barred from the family business**: Ibid., 82.

32 **Years later, Ferdinand Piëch**: Ibid., 26.

CHAPTER 5: CHIEF EXECUTIVE

33 **During the 1950s, Volkswagen**: Markus Lupa, *Volkswagen Chronik: Der Weg zum Global Player* (Wolfsburg: Volkswagen AG, 2008), 46.

34 **An early ad by the New York agency**: Bob Garfield, "Ad Age Advertising Century: The Top 100 Campaigns," *Advertising Age*, March 29, 1999, http://adage.com/article/special-report-the-advertising-century/ad-age-advertising-century-top-100-campaigns/140918/.

34 **Hahn liked Bernbach's lack of pretension**: Interview with Carl Hahn, March 31, 2016.

35 **Volkswagen exported 330,000**: Lupa, *Volkswagen Chronik*, 60.

36 **overtaking Ford's Model T**: Ibid., 90.

36 **In 1961, the West German government**: Steven Parissien, *The Life of the Automobile: The Complete History of the Automobile* (New York: St. Martin's Press, 2014), 124.

37 **despite his close association with Hitler**: Hans Mommsen and Manfred Grieger, *Das Volkswagenwerk und seine Arbeiter im Dritten Reich* (Düsseldorf: ECON Verlag, 1997), 939.

37 **Porsche's attempts to design**: Anton Hunger and Dieter Landenberger, *Porsche Chronicle, 1931–2008* (Munich: Piper Verlag, 2008), 63.

39 **Piëch saw the parent company**: Ferdinand Piëch, with Herbert Völker, *Auto.Biographie* (Hamburg: Hoffmann und Campe Verlag, 2002), 80.

39 **Clean Air Act of 1970**: "Hearings Set on Automobile Pollution Control," Environmental Protection Agency press release, March 4, 1971, http://www.epa.gov/aboutepa/hearings-set-automobile-pollution-control.

39 **The European Union . . . did not have**: "Assessment of the Effectiveness of European Air Quality Measures and Policies, Case Study 2: Comparison

of the US and EU Air Quality Standards & Planning Requirements" (DG Environment, Oct. 4, 2004), http://ec.europa.eu/environment/archives/cafe/activities/pdf/case_study2.pdf, p. 1.

39 **Piëch solved the emissions problem**: Piëch, *Auto.Biographie*, 81.

40 **Behles was sidelined**: Ibid., 86.

41 **who were duly astonished**: Ibid., 105.

41 **By 1984, he had nine children**: Ibid., 126.

43 **he forbade such duplication**: Ibid., 124.

43 **Piëch worried about how to get**: Ibid., 123.

44 **Audi unveiled its first TDI model**: Oliver Strohbach, "Das große Wettbrennen: Mit dem TDI Von Malmö nach Kopenhagen," *Dialoge.Online* (Audi online magazine), http://audi-dialoge.de/magazin/technologie/01-2015/134-das-grosse-wettbrennen.

44 **Piëch was proud of the innovation**: Audi website, "TDI Chronicle," http://www.volkswagenag.com/content/vwcorp/info_center/en/themes/2014/08/Light_my_fire/TDI_chronicle.html.

45 **there was even a TDI Club**: TDI Club website, https://www.tdiclub.com/.

46 **When Böhm celebrated his ninetieth**: "Audi Betriebsrat—Fritz Böhm wird 90 Jahre alt," *AutoNewsBlog*, Dec. 21, 2014, http://www.auto-news-blog.de/audi-betriebsrat-fritz-bohm-wird-90-jahre-alt/.

46 **Hahn commented that Piëch's ascension**: Piëch, *Auto.Biographie*, 135.

46 **The manager found another job**: Ibid., 110.

46 **Piëch saw the man as overly**: Piëch, *Auto.Biographie*, 141.

47 **the weeping family members**: Ibid., 142.

47 **The last—the 21,529,464th**: Lupa, *Volkswagen Chronik*, 188.

48 **Volkswagen had not kept pace:** Piëch, *Auto.Biographie*, 184–87.

48 **Volkswagen lacked early warning systems**: Piëch, *Auto.Biographie*, 147.

48 **Hahn believed that he had been made:** Carl H. Hahn, *Meine Jahre mit Volkswagen* (Munich: Signum, 2005), 289–302.

49 **By Piëch's account**: Ibid., 154–55.

49 **"Only when a company is"**: Ibid., 286 (translation by the author).

CHAPTER 6: BY ALL MEANS NECESSARY

50 **far too many . . . to operate profitably**: Ferdinand Piëch, with Herbert Völker, *Auto.Biographie* (Hamburg: Hoffmann und Campe Verlag, 2002), 156.

51 **For example, female workers**: Christine Holch, "Wie López in den VW-Werken die Arbeit revolutionierte," *Die Zeit*, Dec. 6, 1996.

52 **Under Wendelin Wiedeking**: Nathaniel C. Nash, "Putting Porsche in the Pink," *New York Times*, Jan. 20, 1996, http://www.nytimes.com/1996/01/20/business/putting-porsche-in-the-pink.html?pagewanted=all.

52 **the elimination of 1,850 jobs**: Anton Hunger and Dieter Landenberger, *Porsche Chronicle, 1931–2004* (Munich: Piper Verlag, 2008), 137.

53 **Volkswagen was using a single platform**: Markus Lupa, *Volkswagen Chronik: Der Weg zum Global Player* (Wolfsburg: Volkswagen AG, 2008), 162.

54 **Under Piëch's demanding eye**: Ferdinand Piëch, with Herbert Völker, *Auto.Biographie* (Hamburg: Hoffmann und Campe Verlag, 2002), 239.

55 **It did not take long**: Lupa, *Volkswagen Chronik*, 153.

55 **Even Carl Hahn**: Carl H. Hahn, *Meine Jahre mit Volkswagen* (Munich: Signum, 2005), 42–44.

55 **a typical Volkswagen worker**: "Tarifrunde 1993: Vier-Tage-Woche bei Volkswagen," Wirtschafts- und Sozialwissenschaftliche Institut, http://www.boeckler.de/wsi-tarifarchiv_3267.htm.

55 **the company had thirty thousand**: Lupa, *Volkswagen Chronik*, 162.

57 **"Often codetermination leads"**: Hahn, *Meine Jahre mit Volkswagen*, 294.

58 **"Neither side observed"**: Werner Widuckel, "Paradigmenentwicklung der Mitbestimmung bei Volkswagen," *Schriften zur Unternehmensgeschichte von Volkswagen* (Wolfsburg: Volkswagen AG, 2004), 16.

59 **Before the end of 1994**: "Volkswagen: Piech-Opfer Nummer neun," *Der Spiegel*, Sept. 26, 1994, http://www.spiegel.de/spiegel/print/d-13692675.html.

59 **Hahn pushed for more open debate**: Hahn, *Meine Jahre mit Volkswagen*, 39.

59 **"A climate of fear"**: "Angst und Mißtrauen: Hat der gefeuerte Audi-Chef versagt—oder sein Vorgänger Ferdinand Piech?," *Der Spiegel*, http://www.spiegel.de/spiegel/print/d-13684200.html.

60 **Herman recalled later**: Interview with David Herman, Oct. 19, 2015.

60 **his aides wore their wristwatches**: Keith Bradsher, "Former G.M. Executive Indicted on Charges of Taking Secrets," *New York Times*, May 23, 2000, http://www.nytimes.com/2000/05/23/business/former-gm-executive-indicted-on-charges-of-taking-secrets.html?ref=topics.

61 **Even Piëch, a famously demanding**: Piëch, *Auto.Biographie*, 177.

61 **Piëch was pleased to discover**: Ibid., 161.

61 **promising to consider building**: Ibid., 172.

61 **When rumors began to circulate**: "Kisten nach Amorebieta," *Der Spiegel*, 2 Aug. 1993, pp. 82–86.

61 **According to a lawsuit**: General Motors Corp. v. Ignacio Lopez De Arriortua, 948 F. Supp. 670 (E.D. Mich. 1996), Nov. 26, 1996, http://law.justia.com/cases/federal/district-courts/FSupp/948/670/2098385/.

62 **Piëch denied GM's claims**: Piëch, *Auto.Biographie*, 168.

62 **Piëch later insisted that the word**: Ibid., 175.

62 **"VW set out to destroy Opel"**: Herman interview.

63 **GM sought as much as $4 billion**: Robyn Meredith, "VW Agrees to

Pay G.M. $100 Million in Espionage Suit," *New York Times*, Jan. 10, 1997, http://www.nytimes.com/1997/01/10/business/vw-agrees-to-pay-gm-100-million-in-espionage-suit.html.

63 **Volkswagen did not admit wrongdoing**: Ibid.

63 **The German prosecutors concluded**: Diana T. Kurylko and James R. Crate, "The Lopez Affair," *Automotive News Europe*, Feb. 20, 2006, http://europe.autonews.com/article/20060220/ANE/60310010/the-lopez-affair.

64 **Appearing before the Spanish court**: Emma Daly, "Court Fight in G.M. Spy Case," *New York Times*, May 22, 2001, http://www.nytimes.com/2001/05/22/business/court-fight-in-gm-spy-case.html?ref=topics.

64 **Years later, business schools still used**: Michael H. Moffett and William Youngdahl, *Jose Ignacio Lopez de Arriortua*, Case #: A02-98-0003, Thunderbird School of Global Management, Arizona State University (Glendale, AZ, 1998), http://caseseries.thunderbird.edu/case/jose-ignacio-lopez-de-arriortua.

65 **He could not . . . perceive anything**: Piëch, *Auto.Biographie*, 196.

65 **according to the European Automobile Manufacturers**: Consolidated Registrations by Manufacturer, European Automobile Manufacturers Association, 2016, http://www.acea.be/statistics/article/consolidated-registrations-by-manufacturer.

66 **"If you want to trace"**: Herman interview.

CHAPTER 7: ENFORCERS

68 **"I would scratch my head"**: Interview with Leo Breton, June 4, 2016.

68 **In 1993, EPA testers**: U.S. Justice Department, "U.S. Announces $45 Million Clean Air Settlement with GM" (news release), Nov. 30, 1995, https://www.justice.gov/archive/opa/pr/Pre_96/November95/596.txt.html.

69 **"These so-called defeat devices"**: Ibid.

70 **"The real world could only"**: E-mail from Leo Breton, June 23, 2016.

70 **There was no money for science projects**: The EPA did not respond to several requests for comment on Breton's account of events. Nor did it make officials available for interviews.

71 **The prototype looked like**: Leo Breton, "ROVER: U.S. Environmental Protection Agency's Real-time On-Road Vehicle Emissions Reporter," presentation to the 8th Integer Emissions Summit & DEF Forum USA, Chicago, Oct. 29, 2015.

71 **Ford paid $7.8 million**: "Ford Motor Company Clean Air Act Settlement" (EPA news release), https://www.epa.gov/enforcement/ford-motor-company-clean-air-act-settlement. See also United States of America v. Ford Motor Co., Consent Decree, June 4, 1998, retrieved from https://www.epa.gov/sites/production/files/documents/fordmotor-cd.pdf.

72 **"When he got off the phone"**: Leo Breton, written notes provided to the author, June 2016.

73 **Cummins Engine Co. Inc.**: Consent decree, United States of America v. Cummins Engine Co. Inc., United States District Court for the District of Columbia, 1998 (no precise date given), https://www.epa.gov/sites/production/files/2013-09/documents/cumminscd.pdf.

73 **All told, the settlement cost**: Environmental Protection Agency, "DOJ, EPA Announce One Billion Dollar Settlement with Diesel Engine Industry for Clean Air Violations" (news release), Oct. 22, 1998, https://yosemite.epa.gov/opa/admpress.nsf/b1ab9f485b098972852562e7004dc686/93e9e651adeed6b7852566a60069ad2e?OpenDocument.

74 **But Breton said he was punished**: E-mail from Leo Breton, April 27, 2016.

74 **inventions that generated patent royalties**: United States Patent Number 6,470,732; Leo Alphonse Gerard Breton (inventor), *Real-time exhaust gas modular flowmeter and emissions reporting system for mobile apparatus,* October 29, 2002.

74 **The report portrayed the EPA**: U.S. House of Representatives Committee on Commerce, "Bliley Statement on EPA Diesel Rule Says, 'EPA Enforcement Is All Show and No Go'" (news release), May 17, 2000.

75 **The testing lab in Alexandria**: Peter Whoriskey, "The EPA Closed the Lab That Might Have Caught VW Years Ago," *Washington Post*, Oct. 5, 2015, https://www.washingtonpost.com/business/economy/the-epa-closed-the-lab-that-might-have-caught-vw-years-ago/2015/10/05/03487752-67b5-11e5-9223-70cb36460919_story.html.

75 **By 2002, 40 percent of new**: European Automobile Manufacturers Association, "Share of Diesel in New Passenger Cars," http://www.acea.be/statistics/tag/category/share-of-diesel-in-new-passenger-cars.

76 **In 1998, soon after**: Per Kageson, "Cycle Beating and the EU Test Cycle for Cars," European Federation for Transport and Environment, Nov. 1988, http://www.transportenvironment.org/sites/te/files/media/T&E%2098-3_0.pdf, p. 2.

78 **"The only thing we were missing"**: Interview with Dan Carder, May 24, 2016.

79 **To Greg Thompson, an associate**: Interview with Greg Thompson, May 24, 2016.

79 **Given a bicycle for his birthday**: "WVU's Dan Carder among Time's 100 Most Influential People in the World" (West Virginia University news release), April 21, 2016, http://wvutoday.wvu.edu/n/2016/04/21/wvu-s-dan-carder-among-time-s-100-most-influential-people-in-the-world.

79 **At sixteen, he restored**: Dan Carder, e-mail to author, July 12, 2016.

80 **Carder and Thompson cobbled together**: Gregory J. Thompson, Daniel

K. Carder, Nigel N. Clark, and Mridul Gautam, "Summary of In-use NOx Emissions from Heavy-Duty Diesel Engines," *Journal of Commercial Vehicles* 117 (2009): 162–84.

CHAPTER 8: IMPOSSIBLE DOESN'T EXIST

83 **Bentley had a single profitable year**: Markus Lupa, *Volkswagen Chronik: Der Weg zum Global Player* (Wolfsburg: Volkswagen AG, 2008), 218–22.

83 **Bugatti's founder, Ettore Bugatti**: Ibid., 224–27.

83 **only three were sold**: "1926: Type 41 Royale," http://www.bugatti.com/fr/tradition/histoire/.

83 **Ferruccio Lamborghini was the only**: Lupa, *Volkswagen Chronik*, 228–33.

84 **Some Volkswagen executives**: Interview with Stephan Winkelmann, chief executive of Lamborghini, March 2, 2016.

84 **his interest in Bugatti was inspired**: Ferdinand Piëch, with Herbert Völker, *Auto.Biographie* (Hamburg: Hoffmann und Campe Verlag, 2002), 224.

85 **The company invested 1.1 billion**: Lupa, *Volkswagen Chronik*, 220–23.

85 **Paris Hilton drove one**: Nicholas Schmidle, "The Digital Dirt," *New Yorker*, Feb. 22, 2016.

86 **His aim was to position Bugatti**: Piëch, *Auto.Biographie*, 224.

86 **Piëch put one of his inner circle**: Ibid., 233.

86 **Michelin produced a tire**: Ray Hutton, "Inside Château Bugatti: Piëch's Million-Euro Supercar Inches Carefully Closer to Reality," *Road and Track*, May 2005, http://www.caranddriver.com/features/inside-chateau-bugatti.

86 **The Veyron sold better**: Lupa, *Volkswagen Chronik*, 227.

87 **A Passat driver who wanted to move**: Piëch, *Auto.Biographie*, 265.

88 **"The entire assembly"**: "A Car Factory in the Centre of Town," Volkswagen website, accessed Feb. 27, 2016, https://www.glaesernemanufaktur.de/en/idea.

88 **"Only after sighting the tachometer"**: Peter Robinson, "Volkswagen Phaeton 4MOTION W-12: A People's Car for People with Princely Incomes," *Car and Driver*, Sept. 2002.

88 **The twelve-cylinder motor was so tightly**: Matthias Kriegel, "Produktionsende des VW Phaeton: Nicht mal der Papst konnte ihn retten," *Spiegel Online*, March 10, 2016, http://www.spiegel.de/auto/aktuell/vw-phaeton-selbst-der-papst-konnte-ihn-nicht-retten-a-1078294.html.

89 **By the end of the 1990s**: Volkswagen AG, *Annual Report 1999*, pp. 96–97.

89 **"Then I will tell them"**: Paul A. Eisenstein, "Just Following Orders? Who Was Really Responsible for VW's Emissions Cheating?," *The Detroit Bureau*, Oct. 8, 2015, http://www.thedetroitbureau.com/2015/10/just-following-orders-who-was-really-responsible-for-vws-emissions-cheating.

90 **"I said, 'These body gaps'"**: Interview with Bob Lutz, March 14, 2016. Lutz also told the story during an interview on *Autoline*, Nov. 13, 2015, http://www.autoline.tv/journal/?p=40331. See also Bob Lutz, "One Man Established the Culture That Led to VW's Emissions Scandal," *Road and Track,* Nov. 4, 2015, http://www.roadandtrack.com/car-culture/a27197/bob-lutz-vw-diesel-fiasco/.

90 **"I would not have cared to work for him"**: Interview with Bob Lutz, March 14, 2016.

91 **Asked about these anecdotes**: Conversation with Matthias Prinz on the sidelines of a court hearing in Stuttgart, March 18, 2016.

91 **Ever after, the story goes**: Jack Ewing, "As Boardroom Struggle Ends, Volkswagen Looks to the Future," *New York Times,* April 26, 2015, http://www.nytimes.com/2015/04/27/business/international/as-boardroom-struggle-ends-volkswagen-looks-to-the-future.html.

92 **"Who ordered this car?"**: E-mail from Walter Groth, Dec. 12, 2015.

92 **"in many people he instills fear"**: E-mails from Walter Groth, December 12, 2015, and April 15, 2016.

93 **Arndt Ellinghorst, a management**: Interview with Arndt Ellinghorst, Nov. 2015.

93 **"You have to obey"**: Interview with Arndt Ellinghorst, November 2015.

93 **"The danger with Piech"**: David Woodruff, "Volkswagen's Hard-driving Boss, Ferdinand Piech Is Obsessed with Making Volkswagen into a Global Powerhouse," *BusinessWeek*, Oct. 5, 1996, http://www.businessweek.com/1998/40/b3598015.htm.

94 **The profit in 2002**: Volkswagen AG, *Geschäftsbericht 2002*, 1–2.

94 **The Japanese carmaker earned**: Toyota Motor Corporation, Form 20-F (Annual Report, fiscal year ending March 31, 2003), Securities and Exchange Commission, Washington, DC, 1, 69, http://www.sec.gov/Archives/edgar/data/1094517/000119312503027032/d20f.htm#toc10753_22.

95 **Gold-plated gearshift knobs**: Jack Ewing, "Can Porsche Shine at Volkswagen?," *New York Times*, Aug. 28, 2010.

96 **At the dinner**: The author attended the dinner in 2008 and sat at Wiedeking's table.

96 **Porsche paid €780 million**: *Angebotsunterlage Pflichtangebot der Dr. Ing. h.c. F. Porsche Aktiengesellschaft an die Aktionäre der Volkswagen Aktiengesellschaft* (Prospectus for Porsche offer to buy Volkswagen shares, filed with stock market regulator BaFin), April 26, 2007, 24, http://www.bafin.de/SharedDocs/Downloads/DE/Angebotsunterlage/porsche.pdf?__blob=publicationFile.

98 **Piëch even wore a special pair**: Piëch, *Auto.Biographie*, 282.

98 **Piëch, chilled to the bone**: Ibid., 284.

98 **There was some grumbling**: Henning Krogh, "Gros der Anteilseigner ist

zufrieden mit der Ära Piëch," *Automobilwoche*, April 16, 2002, http://www.automobilwoche.de/article/20020416/NACHRICHTEN/204160702/gros-der-anteilseigner-ist-zufrieden-mit-der-ara-piech-vw-hauptversammlung-ohne-eklat.

99 **The drive in the *Einliterauto***: Piëch, *Auto.Biographie*, 285–86.

99 **He sketched a vision**: Ibid., 286.

CHAPTER 9: LABOR RELATIONS

101 **After a maximum of two years**: Bundesagentur für Arbeit, "Dauer des Anspruchs," accessed May 21, 2016, https://www.arbeitsagentur.de/web/content/DE/BuergerinnenUndBuerger/Arbeitslosigkeit/Arbeitslosengeld/DauerdesAnspruchs/Detail/index.htm?dfContentId=L6019022DSTBAI485667.

102 **She is popular**: Stacey Dooley, "Stacey Dooley at K5 Relax," BBC Three, Oct. 21, 2013, https://www.youtube.com/watch?v=vm6o_2chMgI.

102 **He used the money**: "Lustreisen und ein Lamborghini," *Süddeutsche Zeitung*, Sept. 21, 2010, http://www.sueddeutsche.de/wirtschaft/vw-korruptionsaffaere-lustreisen-und-ein-lamborghini-1.1003074.

103 **He had a parking spot**: Landgericht Braunschweig, decision from Jan. 25, 2007 (case number 6 KLs 48/061).

103 **If the women did not appear**: "Gebauer, wo bleiben die Weiber?," *Die Welt*, June 20, 2015, http://www.welt.de/print/die_welt/wirtschaft/article142799194/Gebauer-wo-bleiben-die-Weiber.html.

104 **"It's about more than just me"**: Peter Hartz, resignation statement, as reprinted verbatim in *Handelsblatt*, July 8, 2005, http://www.handelsblatt.com/unternehmen/industrie/im-wortlaut-dokumentation-erlaerung-von-peter-hartz/2523644.html.

104 **The auditors had found evidence**: "Volkswagen Announces Ombudsman System throughout the Group of Companies" (press release), Nov. 11, 2005, http://www.volkswagenag.com/content/vwcorp/info_center/en/news/2005/11/Volkswagen_Announces_Ombudsman_System_throughout_the_Group_of_Companies.html.

105 **"Anyone who did such a thing"**: Ferdinand Piëch, Zeugenvernehmung, Landeskriminalamt Niedersachsen, Dezernat 36, March 29, 2006. Lawyers for Mr. Piëch did not respond to multiple requests for comment.

105 **"I never distributed"**: The original German reads, "Ich habe niemals Geld verteilt, sondern in diesen unangenehmen Fällen mich dadurch aus der Schlinge gezogen, dass ich es an jemand anderen delegiert habe."

105 **Piëch was called to testify**: "Piëch will nichts gewusst haben," *Zeit Online*, Jan. 3, 2008, http://www.zeit.de/online/2008/02/prozess-vw-affaere-piech.

105 **he arrived at the courthouse**: Kate Connolly, "Bribery, Brothels, Free Viagra: VW Trial Scandalises Germany," *Guardian*, Jan. 13, 2008, https://www.theguardian.com/world/2008/jan/13/germany.automotive.

106 **Hartz admitted in court**: "VW Sex and Bribery Scandal: Sentences Handed Down in Corruption Affair," *Spiegel Online*, Feb. 22, 2008, http://www.spiegel.de/international/business/vw-sex-and-bribery-scandal-sentences-handed-down-in-corruption-affair-a-537137.html.

106 **He served a year and nine months**: "VW-Affäre: Ex-Betriebsratschef Volkert vorzeitig entlassen," *Spiegel Online*, Sept. 2, 2011, http://www.spiegel.de/wirtschaft/unternehmen/vw-affaere-ex-betriebsratschef-volkert-vorzeitig-entlassen-a-784073.html.

107 **The new technology used a single reservoir**: "Common Rail Injection: Advanced Technology for Diesel Engines," Bosch Auto Parts website, http://de.bosch-automotive.com/en/parts_and_accessories/motor_and_sytems/diesel/common_rail_injection/common_rail_diesel_motorsys_parts.

107 **"The market shares over the years"**: Jack Ewing, "Volkswagen Not Alone in Flouting Pollution Limits," *New York Times*, June 9, 2016, http://www.nytimes.com/2016/06/10/business/international/volkswagen-not-alone-in-flouting-pollution-limits.html?ref=topics.

110 **he canceled sales of the Phaeton**: Mark Landler, "VW's Chief, under Fire, Fights Back," *New York Times*, March 8, 2006, http://www.nytimes.com/2006/03/08/automobiles/08volkswagen.html?pagewanted=print&_r=0.

110 **in an interview in March 2006**: Ibid.

CHAPTER 10: THE CHEAT

112 **Martin Winterkorn, not yet a year**: "Offensive 2018," *Der Spiegel*, Nov. 12, 2007, http://www.spiegel.de/spiegel/print/d-53621814.html.

112 **The cars were clean, practical**: "Auf den Punkt: BlueMotion-Offensive," Volkswagen website, Sept. 11, 2007, http://www.volkswagenag.com/content/vwcorp/info_center/de/news/2007/09/to_the_point_bluemotion_offensive.html.

114 **Some engineers at the Wolfsburg**: Interview with a Volkswagen engineer who was knowledgeable about the internal debate but insisted on anonymity.

114 **the urea emissions cleansing system**: Vicente Franco, Francisco Posada Sánchez, John German, and Peter Mock, "Real-World Exhaust Emissions from Modern Diesel Cars," International Council on Clean Transportation, Oct. 2014, p. 7.

114 **Christian Wulff, the prime minister**: Richard Milne, "VW Did Approach DaimlerChrysler over Stake," *Financial Times*, Sept. 30, 2005, https://next.ft.com/content/039a41c8-31a5-11da-9c7f-00000e2511c8.

114 **Because the lean NOx trap**: Volkswagen of America, "Self Study Program 826803 2.0 Liter TDI Common RailvBIN5 ULEV Engine" (service manual), 2008, 24.

116 **"It was impossible to explain"**: Interview with Walter Groth, May 2, 2016.

117 **His parents were ethnic Germans**: Nicola Clark and Melissa Eddy, with reporting by Jack Ewing and Bill Vlasic, "Volkswagen's Chief in the Vortex of the Storm," *New York Times*, Sept. 22, 2015, http://www.nytimes.com/2015/09/23/business/international/volkswagens-chief-in-the-vortex-of-the-storm.html?_r=0.

118 **A former soccer goalkeeper**: Joann Muller, "How Volkswagen Will Rule the World," *Forbes*, May 6, 2013, http://www.forbes.com/sites/joannmuller/2013/04/17/volkswagens-mission-to-dominate-global-auto-industry-gets-noticeably-harder/#120161821ab6.

119 **As Volkswagen later admitted in court**: Göhmann Rechtsanwälte, In Sachen TILP Rechtsanwälte gegen Volkswagen Aktiengesellschaft (Volkswagen law firm's reply to lawsuit filed on behalf of shareholders), Feb. 29, 2016, p. 46.

120 **In November 2006 . . . an engineer**: In re: Volkswagen "Clean Diesel" Marketing, Sales Practices, and Products Liability Litigation, Amended Consolidated Consumer Class Action Complaint, as filed in U.S. District Court, Northern District of California, San Francisco Division, Sept. 2, 2016, p. 152.

121 **The highest ranking executive present**: The author's account of this decisive meeting is based on interviews with several sources with firsthand knowledge of the discussion and with people who have seen documents that were presented at the meeting. The meeting and Krebs' involvement are also described in a documentary and article by Norddeutscher Rundfunk, "VW: Zeugenaussagen belasten früheren Leiter der Motorenentwicklung—vorerst keine Beweise gegen damalige Vorstände," April 25, 2016. Other sources include the class action lawsuit in the United States, In re: Volkswagen "Clean Diesel," Amended Consumer Class Action Complaint filed September 2, 2016, 152–53, which cites documents available to plaintiff's attorneys but not the public, as well as the California State Teachers' Retirement System et al. v. Volkswagen AG complaint filed in state court in Braunschweig, Germany, June 20, 2016, 40, paragraph 52, which cites a Volkswagen employee who is cooperating with German prosecutors.

121 **The presentation was just three pages long**: The PowerPoint was described to the author by three people who have seen it.

122 **described Krebs as being uneasy**: VW Franchise Dealer Amended and Consolidated Class Action Complaint, 32.

122 **Some people had serious reservations**: In re: Volkswagen "Clean Diesel," 153.

122 **chance of getting caught**: United States of America v. Volkswagen AG, Plea Agreement, Statement of Facts, January 11, 2017, Exhibit 2-14. (The

document does not mention Krebs by name but attributes the quote to the head of VW brand engine development, the position that Krebs held.)

122 **there was no technology available**: Göhmann Rechtsanwälte, In Sachen TILP Rechtsanwälte gegen Volkswagen Aktiengesellschaft, 50.

123 **The company at that time paid $120,000**: "Volkswagen to Pay $120,000 to settle complaint by EPA," *The Wall Street Journal,* March 13, 1974, archived by Center for Auto Safety, http://www.autosafety.org/wp-content/uploads/import/VW%20Defeat%20Device%20%24120%2C00%20fine%203-12-74%20Pr.pdf.

123 **Volkswagen paid a $1.1 million penalty**: United States of America v. Volkswagen of America, Consent Decree, United States District Court for the District of Columbia, case 1:05-cv-01193-GK.

123 **Audi engineers had invented**: Lawsuit filed by State of New York, and the New York State Department of Environmental Conservation, by Eric T. Schneiderman, Attorney General of the State of New York v. Volkswagen Aktiengesellschaft d/b/a Volkswagen Group and/or Volkswagen AG; Audi AG; Volkswagen Group of America, Inc.; Dr. Ing. H.C. F. Porsche AG d/b/a, Porsche AG; and Porsche Cars North America, Inc., July 19, 2016, retrieved from https://consumermediallc.files.wordpress.com/2016/07/new-york-vw-complaint-7-19.pdf, pp. 21–22.

123 **So Audi developed software**: State of New York v. Volkswagen et. al., 21–22, as well as in the complaint filed in Maryland Department of the Environment v. Volkswagen Aktiengesellschaft d/b/a Volkswagen Group and/or Volkswagen AG; Audi AG; Volkswagen Group of America, Inc.; Dr. Ing. h.c.f. Porsche AG d/b/a, Porsche AG; and Porsche Cars North America, Inc., July 19, 2016, 26–27. Commonwealth of Massachusetts, v. Volkswagen AG; Audi AG; Volkswagen Group of America, Inc.; Dr. Ing. H.C. F. Porsche AG, d/b/a Porsche AG; and Porsche Cars North America, Inc., 22. The German newspaper Handelsblatt reported the existence of the Audi acoustic function (see "Exclusive: VW's Diesel Manipulation Software Originated at Luxury Car Subsidiary Audi, Sources Say," Handelsblatt, April 20, 2016, https://global.handelsblatt.com/breaking/exclusive-vws-diesel-manipulation-software-originated-at-luxury-car-subsidiary-audi-sources-say), although that article said that the device was not deployed until VW began to install it.

124 **Bosch advertised that the new computer**: "The Brain of Diesel Injection: New Bosch EDC17 Engine Management System" (Bosch press release), Feb. 2006.

124 **Bosch's new EDC 17**: In re: Volkswagen "Clean Diesel" Marketing, Sales Practices, and Products Liability Litigation, Defendants Volkswagen AG et al. Motion to Dismiss the Consolidated Securities Class Action Complaint; Memorandum of Law in Support Thereof, 28. (In its defense against a suit by shareholders,

Volkswagen referred to Bosch's role as producer of the software code as part of its argument that Winterkorn would not have known about the defeat device.)

124 **Bosch worked closely with Volkswagen**: In re: Volkswagen "Clean Diesel," Consolidated Consumer Class Action Complaint, 145–52.

124 **In June 2008, a Bosch executive**: The text of Bosch's letter to Volkswagen asking to be indemnified is quoted verbatim in a complaint filed against Volkswagen on January 3, 2017, in the Landgericht Braunschweig by the law firm Hausfeld Rechtsanwälte on behalf of an anonymous Volkswagen owner, pages 10–12. The existence of a disclaimer that would exclude Bosch from liability was alluded to in testimony before the European Parliament. See: Committee of Inquiry into Emission Measurements in the Automotive Sector (EMIS), Hearing of representatives of Robert BOSCH GmbH, September 15, 2016, 15. https://polcms.secure.europarl.europa.eu/cmsdata/upload/161c2028-a174-4776-a20b-5507d4fdb51c/CRE_EMIS_15%2009%202016_EN_redacted.pdf. (A Bosch representative declined to comment in response to a question from a member of Parliament and said he was not aware of any disclaimer.)

124 **"yet another path"**: Text as quoted in Hausfeld complaint against Volkswagen, 10.

124 **a Volkswagen executive in engine development**: Ibid., 155.

124 **referred to a different group of engines**: In re: Volkswagen "Clean Diesel" Marketing, Sales Practices, And Products Liability Litigation, Plaintiffs' Memorandum in Support of Preliminary Approval of the Bosch Class Action Settlement Agreement, January 31, 2017, 13.

124 **As Volkswagen later admitted**: Jack Ewing, "VW Says Emissions Cheating Was Not a One-time Error," *New York Times*, Dec. 10, 2015, http://www.nytimes.com/2015/12/11/business/international/vw-emissions-scandal.html.

125 **In November 2006, Stuart Johnson**: State of New York v. Volkswagen AG, 30. The relevant passage in the New York complaint reads in full: "He [Johnson] referenced an earlier lawsuit in which heavy-duty engine manufacturers were caught using 'cycle beating strategies [with] timers on them that enacted the injection timing change once the engine was in a mode for a specific length of time' as a 'clear violation of the spirit of the emission regulations and the certification test procedure.' " The same language is cited in Maryland Department of the Environment v. Volkswagen AG et al., 35–36; and Commonwealth of Massachusetts v. Volkswagen AG et al., 28–29.

125 **A few days later, Leonard Kata**: New York v. Volkswagen, 31; Maryland v. Volkswagen, 36–37; Massachusetts v. Volkswagen, 29–30. According to the complaints, Kata wrote: "In connection with the introduction of future diesel products, there has been considerable discussion recently regarding the identification of Auxiliary Emission Control Devices (AECDs). . . . The agencies' interest in the identification of AECDs is to determine whether any of these devices can

be considered a defeat device." Kata also wrote, according to the complaints: "EPA also discusses the concept of the existence of a defeat device strategy if a manufacturer's choice of basic design strategy cannot provide the same degree of emission control during both [emissions-test cycle] and [non-emissions-test cycle] operation when compared with other systems available in the industry."

125 **Anything that caused the car**: Johnson and Kata did not respond to requests for comment.

126 **"Volkswagen agrees"**: State of New York v. Volkswagen AG, 32, Maryland v. Volkswagen, 37; Massachusetts v. Volkswagen, 30. The text quoted in the complaints reads: "VW [*sic*] position regarding 'normal vehicle operation' is that the light-duty vehicle emission test procedures cover normal vehicle operation in customer's hands. [CARB official] Duc Nguyen expects emission control systems to work during conditions outside of the emissions tests. Volkswagen agrees." According to Alberto Ayala, deputy executive officer of the California Air Resources Board, Volkswagen was told on "multiple occasions" that emissions systems had to function outside the lab (e-mail from Ayala, December 31, 2016).

126 **In 2007, Wolfgang Hatz**: Danny Hakim and Jack Ewing, "VW Executive Had a Pivotal Role as Car Maker Struggled with Emissions," *New York Times*, Dec. 21, 2015, http://www.nytimes.com/2015/12/22/business/international/vw-executive-had-a-pivotal-role-as-car-maker-struggled-with-emissions.html.

126 **"CARB is not realistic"**: "Dr. Wolfgang Hatz talks at German Tecday about CARB and CAFE," DrivingTheNation.com, retrieved from https://www.youtube.com/watch?v=zThmune955g.

127 **"We don't need BlueTec"**: Mark Stevens, "VW, Audi Will End Joint US Diesel Campaign with D/C," *Automotive News Europe*, Aug. 8, 2007, http://www.autonews.com/article/20070808/ANE01/70808010/vw-audi-will-end-joint-us-diesel-campaign-with-d/c.

127 **Instead, the Q7s were also programmed**: Volkswagen Plea Agreement, Exhibit 2-15. In the plea agreement, Volkswagen admits that cars with three-liter diesel motors were equipped with defeat devices designed to help conserve consumption of AdBlue diesel exhaust fluid. See also: State of New York vs. Volkswagen AG et al. (lawsuit filed by New York Attorney General Eric T. Schneiderman), July 19, 2016, 24. See also In re: Volkswagen "Clean Diesel" Marketing, Sales Practices, and Products Liability Litigation, Partial Consent Decree with U.S. government, December 20, 2016, https://www.epa.gov/sites/production/files/2016-12/documents/30literpartialconsentdecree.pdf. In the partial consent decree, Audi admits that vehicles with three-liter motors from model years 2009 to 2015 contained defeat devices. Volkswagen on behalf of Audi has agreed to a $1 billion settlement covering the three-liter motors. There is a three or four-year lead time from when design decisions are made and the cars come on the market, so it would be logical

that decisions made in 2006 would apply to 2009 model cars, which came on the market in late 2008.

127 **Winterkorn . . . was informed**: Ibid.

127 **But documents indicate that the decision**: Ibid., 24–25.

128 **"The steep pace of mobility growth"**: Martin Winterkorn, text of closing remarks to Vienna Motor Symposium, April 25, 2008, as reprinted in "Conference Report: 29th International Vienna Motor Symposium," *MTZextra*, 31.

CHAPTER 11: THE PORSCHES AND THE PIËCHS

130 **Reporters were scolded**: The author is among those who have been scolded.

130 **In its annual report, Volkswagen**: Volkswagen AG, *Annual Report 2004*, 38.

132 **"Porsche's strategic objective"**: Anton Hunger and Dieter Landenberger, *Porsche Chronicle, 1931–2004* (Munich: Piper Verlag, 2008), 168.

132 **"For Porsche as a niche provider"**: Wendelin Wiedeking, statement to State Court in Stuttgart, Oct. 22, 2015, p. 1 (not publicly available).

134 **the two men began to argue bitterly**: Ibid., 10.

134 **Wiedeking spoke dismissively**: Dietmar H. Lamparter, "Macht ist ihm wichtiger als Geld: Warum sich VW-Aufsichtsrat Ferdinand Piëch gegen Porsche und damit gegen die eigene Familie stellt," *Die Zeit*, Sept. 18, 2008, http://www.zeit.de/2008/39/Piech-Porsche.

134 ***Die Zeit*, an influential weekly**: Ibid.

134 **"an ordinary manager"**: Piëch used the German word *Angestellter*, meaning a salaried employee. The word implies ordinariness.

134 **On October 23, 2007**: European Commission press release No. 74/07, Judgment of the Court of Justice in Case C-112/05, Commission of the European Communities v Federal Republic of Germany, *The Volkswagen Law Restricts the Free Movement of Capital*, Oct. 23, 2007.

135 **Wiedeking later described the atmosphere**: Wiedeking, court statement, 14.

135 **Wiedeking later maintained**: Ibid., 15–16.

136 **It was not until September 16, 2008**: Porsche SE, *Annual Report 2007/2008*, 18.

136 **Despite the growing market turbulence**: Transcript of testimony before Stuttgart State Court on Jan. 15, 2016 (not publicly available).

138 **The more Volkswagen shares fell**: U.S. Court of Appeals for the Second Circuit, Parkcentral Global Hub Limited, et al. v. Porsche Automobile Holdings SE, F/K/A Dr. Ing. H.C. F. Porsche AG, Wendelin Wiedeking, Holger P. Härter, Aug. 15, 2014.

138 **on October 21, for example, Maple Bank**: Indictment of Wendelin Wiedeking and Holger Härter by Stuttgart state's attorney (not publicly available).

138 **The strike price on options**: Transcript of final argument by Heiko

Wagenpfeil, Stuttgart state's attorney, in trial of Wendelin Wiedeking and Holger Härter, Feb. 18, 2016 (not publicly available).

139 **"Porsche has decided to make"**: Porsche Automobil Holding SE, "Porsche Heads for Domination Agreement, Interest in Volkswagen Increased to 42.6 Percent" (news release), http://www.porsche-se.com/pho/en/press/newsarchive2008/?pool=pho%20%20&id=2008-10-26.

140 **Another . . . was Adolf Merkle**: Carter Dougherty, "Town Mourns Typical Businessman Who Took Atypical Risks," *New York Times*, Jan. 12, 2009, http://www.nytimes.com/2009/01/13/business/worldbusiness/13merckle.html?_r=0.

142 **"The Porsche myth lives"**: "Wendeling Wiedeking: Tränen beim Abschied," *Süddeutsche Zeitung*, May 17, 2010, http://www.sueddeutsche.de/wirtschaft/wendelin-wiedeking-traenen-beim-abschied-1.168179.

142 **As Richard W. Painter, a professor**: Interview with Richard W. Painter, Feb. 2, 2016.

143 **The deal "was designed to suit"**: Anne Kvam and Ola Peter Krohn Gjessing, Norges Bank Investment Management, letter to Ferdinand Piëch and members of the Volkswagen supervisory board, Oct. 7, 2009, http://www.nbim.no/globalassets/documents/news/2009/2009-10-07_nbim_letter_volkswagen.pdf.

CHAPTER 12: CLEAN DIESEL

146 **The green marketing push**: In re: Volkswagen "Clean Diesel" Marketing, Sales Practices, and Products Liability Litigation, Amended Consolidated Consumer Class Action Complaint, U.S. District Court, Northern District of California, San Francisco Division, Aug. 16, 2016, pp. 166–82

147 **Cynthia Mackie, of Silver Spring**: E-mail from Cynthia Mackey, Nov. 9, 2015.

147 **The Volkswagen buyers were often**: In re: Volkswagen "Clean Diesel," 6–125.

147 **"You can't go wrong with clean diesel"**: Ibid., 49.

147 **TDI exhaust was so clean it "is almost"**: Ibid., 34.

148 **In 2010, for example, a Volkswagen**: Ibid., 175.

148 **Tony German, a health care**: Interview with Tony German, Sept. 21, 2015.

148 **closed a factory in New Stanton**: John Holusha, "Volkswagen to Shut U.S. Plant," *New York Times*, Nov. 21, 1987, http://www.nytimes.com/1987/11/21/business/volkswagen-to-shut-us-plant.html.

149 **"An auto plant is the holy grail"**: Jack Ewing, "Investing in America: Volkswagen Rolls the Dice on Tennessee," *BusinessWeek*, as reprinted in

Spiegel Online, Dec. 11, 2008, http://www.spiegel.de/international/business/investing-in-america-volkswagen-rolls-the-dice-on-tennessee-a-595770.html.

149 **he traveled seven times to the United States**: Declaration of Martin Winterkorn, U.S. Judicial Panel on Multidistrict Litigation, In re: Volkswagen "Clean Diesel" Marketing, Sales Practices, and Products Liability Litigation, MDL No. 2672, Document 1707-1, filed Aug. 1, 2016.

150 **the proportion was much higher**: Volkswagen PowerPoint presentation, "TDI: U.S. Market Success," March 2015.

150 **Diesel cars were also more profitable**: VW Franchise Dealer Amended and Consolidated Class Action Complaint, Case No. 02672-CRB (JSC), Sept. 30, 2016, p. 69.

150 **At the christening of the Chattanooga plant**: "Volkswagen Inaugurates New Plant at Chattanooga/U.S." (Volkswagen press release), May 24, 2011, http://www.volkswagenag.com/content/vwcorp/info_center/en/themes/2011/05/Volkswagen_inaugurates_new_plant_at_Chattanooga_U_S_.html.

150 **The core goal was to sell**: Volkswagen AG, *Annual Report 2009* (online version), http://annualreport2009.volkswagenag.com/managementreport/reportonexpecteddevelopments/strategy/strategy2018.html.

151 **The code acknowledged a duty**: The Volkswagen Group Code of Conduct, http://en.volkswagen.com/content/medialib/vwd4/de/Volkswagen/Nachhaltigkeit/service/download/corporate_governance/Code_of_Conduct/_jcr_content/renditions/rendition.file/the-volkswagen-group-code-of-conduct.pdf, p. 4.

151 **"Stretch goals are very useful"**: Interview with David Bach, June 17, 2016.

152 **Volkswagen engineers drew on the precedent**: In re: Volkswagen "Clean Diesel" Marketing, Sales Practices, and Products Liability Litigation, Partial Consent Decree, December 20, 2016, 3.

153 **Neusser's response . . . was to tell the engineers**: Volkswagen Plea Agreement, Exhibit 2-18, 2-19. Also: United States v. Richard Dorenkamp, Heinz-Jakob Neusser et al., Second Superseding Indictment, January 11, 2017, 19–21.

154 **"Thanks to state-of-the-art engineering"**: Heinz-Jakob Neusser, Jörn Kahrstedt, Richard Dorenkamp, and Hanno Jelden, "The Euro 6 Engines in the Modular Diesel Engine System of Volkswagen," *MTZ—Motortechnische Zeitschrift*, June 2016, pp. 4–10.

154 **"I was beating up on 'em"**: Interview with Bob Lutz, March 14, 2016.

155 **"They said, 'But we're clean!'"**: Jack Ewing and Graham Bowley, "The Engineering of Volkswagen's Aggressive Ambition," *New York Times*, Dec. 13, 2015.

156 **"It doesn't clank"**: "IAA 2011 Hyundai New Generation i30 and

Martin Winterkorn," YouTube, Sept. 15, 2011, https://www.youtube.com/watch?v=YpPNVSQmR5c.

157 **"This clever woman has achieved"**: Mark Christian Schneider, "Die Piëchs bei der VW-Party: Szenen einer fast normalen Ehe," *Handelsblatt*, Oct. 1, 2010, http://www.handelsblatt.com/unternehmen/industrie/die-piechs-bei-der-vw-party-szenen-einer-fast-normalen-ehe-seite-2/3552522-2.html.

158 **Millions had been produced**: Audi presentation reproduced by *Handelsblatt* online, http://www.handelsblatt.com/images/auszug-aus-einer-audi-praesentation/12353470/4-formatOriginal.png.

158 **Volkswagen had deployed defeat devices**: In re: Volkswagen "Clean Diesel," Partial Consent Decree with U.S. government, September 30, 2016, 2–3; Partial Consent Decree with U.S. government, December 20, 2016, 3. https://www.epa.gov/enforcement/30l-second-partial-and-20l-partial-and-amended-consent-decree

158 **the company had programmed the onboard**: In re: Volkswagen "Clean Diesel," Consolidated Consumer Class Action Complaint, September 2, 2016, 156.

CHAPTER 13: ENFORCERS II

159 **There is overwhelming scientific evidence**: "Nitrogen Dioxide (NO2) Pollution," EPA website, https://www.epa.gov/no2-pollution.

160 **When inhaled, nitrogen dioxide**: Environmental Protection Agency, "Integrated Science Assessment for Oxides of Nitrogen—Health Criteria," Jan. 2016, lxxxvii, 1-6 to 1-36.

160 **Studies have found**: EPA, "Integrated Science Assessment."

160 **which also causes asthma**: "Ozone Basics," EPA website, https://www.epa.gov/ozone-pollution/ozone-basics#effects.

160 **contributes to global warming**: National Aeronautics and Space Administration, Goddard Institute for Space Studies, "Tango in the Atmosphere: Ozone and Climate Change," http://www.giss.nasa.gov/research/features/200402_tango/.

161 **Pound for pound, nitrous oxide**: "Overview of Greenhouse Gases," EPA website, https://www.epa.gov/ghgemissions/overview-greenhouse-gases#nitrous-oxide.

161 **tailpipe emissions account for**: EPA, "Integrated Science Assessment," 1-7 and 1-48.

161 **Air pollution of all kinds was responsible**: European Commission Impact Assessment, "Communication from the Commission to the Council, the European Parliament, the European Economic and Social Committee and the Committee of the Regions—A Clean Air Programme for Europe," 2013, pp. 10, 105.

162 **Early in his career**: Axel Friedrich, Volkhard Möcker, and Karl G. Tempel, *Was sie schon immer über Luftreinhaltung wissen wollten* (Berlin: Bundesminister der Innern, 1983).

162 **In 2003, Friedrich's department published**: Umweltbundesamt, "Sachstandspapier: Erhöhte NO_x-Emissionen von EURO 2-Lkw," https://www.umweltbundesamt.de/sites/default/files/medien/publikation/long/3590.pdf.

163 **The German magazine *Stern* described**: Dirk Vincken, "Axel Friedrich: Schadstoff-Ritter in der Abstellkammer," *Stern*, Dec. 7, 2007, http://www.stern.de/auto/service/axel-friedrich-schadstoff-ritter-in-der-abstellkammer-3221498.html.

163 **"I'm a bad loser"**: Interview with Axel Friedrich, April 13, 2016.

163 **Cars are the country's biggest export**: Statistisches Bundesamt, "Wichtigstes deutsches Exportgut 2015: Kraftfahrzeuge," https://www.destatis.de/DE/ZahlenFakten/GesamtwirtschaftUmwelt/Aussenhandel/Handelswaren/Aktuell.Html.

163 **employ about 800,000 people**: Verband der Autoindustrie, "Zahlen und Daten," https://www.vda.de/de/services/zahlen-und-daten/zahlen-und-daten-uebersicht.html.

163 **Wissmann addressed Merkel**: Letter from Matthias Wissmann to Angela Merkel, May 8, 2013. Reproduced at https://www.greenpeace.de/sites/www.greenpeace.de/files/publications/20130508-vda-brief-merkel.pdf.

164 **After Sigmar Gabriel**: Florian Harms, "Politiker-Gehälteraffäre: Auch Sigmar Gabriel stand geschäftlich in Beziehung zu VW," *Der Spiegel*, Feb. 3, 2005, http://www.spiegel.de/politik/deutschland/politiker-gehaelteraffaere-auch-sigmar-gabriel-stand-geschaeftlich-in-beziehung-zu-vw-a-340083.html.

164 **Wissmann wrote a letter**: Matthias Wissmann, Verband der Automobilindustrie, letter to Chancellor Angela Merkel, May 8, 2013.

165 **"We really tried to do"**: Interview with John German, Sept. 2, 2016.

CHAPTER 14: ON THE ROAD

166 **Carder asked Hemanth Kappanna**: Interview with Hemanth Kappanna, June 30, 2016.

167 **"We never set out"**: Interview with Dan Carder, June 30, 2016.

168 **The state has 25 million vehicles**: "History of Air Resources Board," CARB website, http://www.arb.ca.gov/knowzone/history.htm.

168 **Six of the ten U.S. cities**: "Most Polluted Cities," American Lung Association website, http://www.lung.org/our-initiatives/healthy-air/sota/city-rankings/most-polluted-cities.html.

168 **Alberto Ayala, the head**: Interview with Alberto Ayala, Oct. 10, 2015.

168 **Ayala, who has a doctorate**: "Alberto Ayala Deputy Executive Officer

California Air Resources Board," CARB website, https://www.arb.ca.gov/html/org/eo-bios/bios/alberto-ayala.htm.

169 **Arie J. Haagen-Smit, the Dutch**: "History of Air Resources Board."

170 **"an 'aha' moment"**: Interview with Dan Carder, May 24, 2016.

171 **Besch had studied automotive**: Interview with Marc Besch, May 24, 2016.

171 **Thiruvengadam had earned**: Curriculum vitae, Arvind Thiruvengadam, Research Assistant Professor, West Virginia University.

172 **Greg Thompson . . . was pleased**: Interview with Gregory J. Thompson, May 24, 2016.

172 **The 117-page research paper**: Gregory J. Thompson, Daniel K. Carder, Marc C. Besch, Arvind Thiruvengadam, Hemanth K. Kappanna, "In-Use Emissions Testing of Light-Duty Diesel Vehicles in the United States," Center for Alternative Fuels, Engines & Emissions, Department of Mechanical & Aerospace Engineering, West Virginia University, May 15, 2014.

173 **At least one person already suspected**: Interview with John German, Sept. 2, 2016.

174 **Several requested copies of the study**: State of New York v. Volkswagen AG et al. (lawsuit filed by New York Attorney General Eric T. Schneiderman), July 19, 2016, https://consumermediallc.files.wordpress.com/2016/07/new-york-vw-complaint-7-19.pdf, p. 36. Also: State of Maryland v. Volkswagen, 43; Commonwealth of Massachusetts v. Volkswagen, 34.

CHAPTER 15: EXPOSURE

176 **reported directly to Winterkorn**: Declaration of Martin Winterkorn, filed as an exhibit in Winterkorn's Motion to Dismiss the Consolidated Securities Class Action Complaint, August 1, 2016.

176 **Gottweis was among the first**: State of New York v. Volkswagen AG et al. (lawsuit filed by New York Attorney General Eric T. Schneiderman), July 19, 2016, https://consumermediallc.files.wordpress.com/2016/07/new-york-vw-complaint-7-19.pdf, p. 36.

176 **included in the packet**: Cover letter from Frank Tuch, head of group quality assurance, to Martin Winterkorn, May 23, 2014, filed as an exhibit in Volkswagen's Motion to Dismiss the Consolidated Securities Class Action Complaint, August 1, 2016. The report that Gottweis wrote will be referred to below as the "Gottweis memo."

177 **"A thorough explanation"**: Bernd Gottweis, "Notiz an Herrn Tuch," May 23, 2014 (memo to Frank Tuch, Volkswagen head of quality control, which Tuch forwarded to Winterkorn on May 23, 2014). Translation was provided by Volkswagen in the course of court proceedings.

177 **no proof that Winterkorn . . . read the memo**: In re: Volkswagen "Clean Diesel," Motion to Dismiss the Consolidated Securities Class Action Complaint, August 1, 2016, 32.

178 **According to a document . . . the attendees included**: In Re: Volkswagen "Clean Diesel," Amended Consolidated Consumer Class Action Complaint, September 2, 2016, 164–65; http://www.wsj.com/articles/volkswagen-robert-bosch-met-in-2014-to-discuss-emissions-software-suit-says-1473370004 (refers to "the meeting agenda"); https://www.ft.com/content/6e0ec870-798e-11e6-97ae-647294649b28 (refers to "notes from a May 2014 meeting").

178 **a general discussion about diesel engine acoustics**: In re: Volkswagen "Clean Diesel," Plaintiffs' Memorandum in Support of the Bosch Class Action Settlement Agreement, 13.

178 **The proposals were rejected**: Jack Ewing, "VW Presentation in '06 Showed How to Foil Emissions Tests," *New York Times*, April 26, 2016, http://www.nytimes.com/2016/04/27/business/international/vw-presentation-in-06-showed-how-to-foil-emissions-tests.html. The information in the article was based on two sources who attended meetings where the proposals were discussed, but who insisted on anonymity.

179 **An internal Volkswagen presentation**: Kayhan Özgenc and, Jan Wehmeyer, "Skandal-Akte VW: Die ganze Wahrheit über den Abgas-Betrug," *Bild am Sonntag*, February 14, 2016. The author has reviewed the documents upon which the *Bild* report is based.

179 **the emissions issue was a technical problem**: Göhmann Rechtsanwälte, In Sachen TILP Rechtsanwälte gegen Volkswagen Aktiengesellschaft, 56–57. The author has also reviewed Volkswagen internal documents from the period in question.

179 **"It should first be decided"**: United States v. Oliver Schmidt, Criminal Complaint, Affidavit by Ian M. Dinsmore, FBI special agent, December 30, 2016, pages 11–12

180 **In June, Volkswagen quietly tweaked**: State of New York v. Volkswagen AG, 39. Also: State of Maryland v. Volkswagen, 47; Commonwealth of Massachusetts v. Volkswagen, 37. The change to the application for certification was confirmed by Alberto Ayala, deputy executive officer of CARB (e-mail from Ayala, January 5, 2017).

180 **"significant rejection reason"**: State of New York v. Volkswagen AG, 39–40; State of Maryland v. Volkswagen, 47; Commonwealth of Massachusetts v. Volkswagen, 37. See also: Göhmann Rechtsanwälte, In Sachen TILP Rechtsanwälte gegen Volkswagen Aktiengesellschaft, 49, a brief filed on behalf of Volkswagen in response to a lawsuit by shareholders in Germany. The brief notes that one purpose of the defeat device was to allow reduced consumption of the urea solution.

181 **even with the increased urea consumption**: Gottweis memo. Volkswagen

was never able to find a technical solution for the vehicles with illegal software that brought all of the cars into compliance with U.S. standards.

181 **At that point, CARB officials say**: Interview with Stanley Young, CARB spokesman, July 5, 2016.

181 **an unadorned conference room**: Based on author's observations during a tour of the offices and labs on Oct. 11, 2016.

182 **While repeating explanations that officials**: State of New York v. Volkswagen AG, 42.

182 **The update would "optimize" performance**: Volkswagen of America recall notice, April 2015, reproduced in In re: Volkswagen "Clean Diesel," Volkswagen-Branded Franchise Dealer Amended and Consolidated Class Action Complaint, September 30, 2016, 50. (The recall notice is reproduced in the complaint.)

182 **the recall would cure the excess emissions**: Letter from CARB to Volkswagen AG, Audi AG and Volkswagen of America Re: Admission of Defeat Device and California Air Resources Board requests, September 18, 2015. https://www.arb.ca.gov/newsrel/in_use_compliance_letter.pdf. The letter retroactively revoked approval for the recall based on CARB's conclusion that the recall did not fix the emissions problem. See also Gottweis memo, discussed above at page 176.

183 **Volkswagen told CARB officials**: Ibid., 41–43.

183 **"ongoing commitment to our environment"**: Volkswagen of America recall notice, April 2015.

183 **updating the software in 280,000**: Göhmann Rechtsanwälte, In Sachen TILP Rechtsanwälte gegen Volkswagen Aktiengesellschaft (Volkswagen law firm's reply to lawsuit filed on behalf of shareholders), Feb. 29, 2016, p. 53.

183 **but the upgrade did not remove the illegal code**: United States of America v. James Robert Liang, 6.

183 **In fact, Volkswagen brazenly used**: The flash action added functionality that caused the cars to go into good behavior mode when the wheels were moving but the steering wheel remained stationary, as would be the case during a lab test on rollers. E-mail from Felix Domke, software expert who analyzed the Volkswagen engine computer before and after the recall, April 20, 2016. His conclusion was reported by German media, for example: http://www.ndr.de/der_ndr/presse/mitteilungen/Der-zweite-Betrug-VW-Mitarbeiter-erweiterten-illegale-Abschaltvorrichtung-offenbar-noch-2015,pressemeldungndr16946.html.

183 **the steering wheel remained stationary**: Ibid.

184 **In fact, Audi had done its own**: State of New York v. Volkswagen AG, 43–44; State of Maryland v. Volkswagen, 53; Commonwealth of Massachusetts v. Volkswagen, 41. E-mail from Alberto Ayala, January 3, 2017.

184 **"We heard all kinds of explanations"**: E-mail from Alberto Ayala, January 3, 2017.

184 **After the Volkswagen recall was complete**: Ibid., 44–45.

185 **The engineer expressed concern**: State of New York v. Volkswagen AG, 45; State of Maryland v. Volkswagen, 55; Commonwealth of Massachusetts v. Volkswagen, 42. See also: United States of America v. James Robert Liang, grand jury indictment, June 1, 2016, 21. The German engineer's comment is translated in the indictment as, "We need a story for the situation!" The anecdote is also recounted in: Aruna Viswanatha and Mike Spector, "VW Emissions Cheating Ran Deep and Wide, State Alleges," *Wall Street Journal,* July 19, 2016.

185 **an unusual level of scrutiny**: In re: Volkswagen "Clean Diesel," Volkswagen-Branded Franchise Dealer complaint, 34–35.

185 **"Please be aware that this type"**: State of New York v. Volkswagen AG, 46.

CHAPTER 16: PIËCH'S FALL

186 **"Ich bin auf Distanz zu Winterkorn"**: Dietmar Hawranek, "Aufsichtsräte attackieren VW-Boss 'Ich bin auf Distanz zu Winterkorn,'" *Spiegel Online*, April 10, 2015.

186 **"was the best" man for the job**: Stefan Anker, "Ferdinand Piëch attackiert Porsche-Chef Wiedeking," *Welt*, Nov. 17, 2011, https://www.welt.de/wirtschaft/article3723072/Ferdinand-Piech-attackiert-Porsche-Chef-Wiedeking.html.

187 **"the most successful automobile manager"**: Volkswagen Konzernbetriebsrat, Statement von Bernd Osterloh, April 10, 2015.

188 **The company's profit margin**: Volkswagen AG, *Half-Yearly Financial Report January–June 2015*, p. 21.

189 **If Piëch was aware of the emissions**: "Piëch sprach Winterkorn im März 2015 auf Abgasermittlung an," *Bild*, Aug. 28, 2016, http://www.bild.de/geld/aktuelles/wirtschaft/bams-piech-sprach-winterkorn-im-maerz-2015-47538026.bild.html.

189 **Some in the German news media**: David Böcking, "Piëch vs. Winterkorn: Zukunft im Zeichen der Guillotine," *Spiegel Online*, April 13, 2015, http://www.spiegel.de/wirtschaft/unternehmen/ferdinand-piech-vs-martin-winterkorn-was-wird-aus-vw-a-1028296.html.

189 **his mere appearance sparked speculation**: Markus Voss, "Erstaunliches Comeback: Der Alte ist zurück! Wieviel Piëch steckt jetzt wieder in Volkswagen?," *FOCUS-Online*, Sept. 28, 2015, http://www.focus.de/finanzen/boerse/aktien/gut-getimtes-comeback-der-alte-ist-zurueck-wieviel-piech-steckt-jetzt-wieder-in-volkswagen_id_4977616.html.

190 **"It is no exaggeration to say"**: Jack Ewing, "Volkswagen's Chairman, Ferdinand Piëch, Is Ousted in Power Struggle," *New York Times*, April 25, 2015,

http://www.nytimes.com/2015/04/26/business/international/volkswagens-chairman-ferdinand-piech-is-ousted-in-power-struggle.html.

190 **Volkswagen and its related brands**: European Automobile Manufacturers Association, "New Passenger Car Registrations in the European Union," June 16, 2015, http://www.acea.be/uploads/press_releases_files/20150616_PRPC_1505_FINAL.pdf.

CHAPTER 17: CONFESSION

192 **In June . . . CARB tested a 2012 Passat**: State of New York v. Volkswagen AG et al. (lawsuit filed by New York Attorney General Eric T. Schneiderman), July 19, 2016, https://consumermediallc.files.wordpress.com/2016/07/new-york-vw-complaint-7-19.pdf, pp. 45–46.

192 **"We must be sure to prevent"**: United States of America v. James Robert Liang, grand jury indictment, 21.

193 **Among those Horn supposedly informed**: State of New York v. Volkswagen AG, 46, State of Maryland v. Volkswagen, 57; Commonwealth of Massachusetts v. Volkswagen, 44.

193 **no proof that Horn was aware**: In re: Volkswagen "Clean Diesel" Marketing, Defendants Motion to Dismiss, 37. In testimony before Congress, Horn said he was told in the spring of 2014 about "a possible emissions non-compliance," but that he was assured engineers were working with regulators to resolve the issue. See Testimony of Michael Horn, President and CEO of Volkswagen Group of America, Inc. Before the House Committee on Energy and Commerce Subcommittee on Oversight and Investigations October 8, 2015.

193 **The committee's conclusions were summarized**: "Notiz an Herrn Tuch, Q2 2015 TREAD Meeting am 21.07.2015" (Volkswagen internal memo).

194 **were in all likelihood informed**: Göhmann Rechtsanwälte, In Sachen TILP Rechtsanwälte gegen Volkswagen Aktiengesellschaft, 63.

194 **Winterkorn was not informed at the meeting**: Ibid.

194 **Winterkorn later said under oath**: Alison Smale and Jack Ewing, "Ex-Chief of V.W., Testifying in Germany, Stands His Ground on Emissions Deception," Martin Winterkorn testimony to Bundestag Untersuchungsausschuss zur VW-Abgasaffäre (Parliamentary Investigative Committee on the Volkswagen Emissions Affair), January 19, 2017.

194 **One of Volkswagen's in-house lawyers**: Göhmann Rechtsanwälte, In Sachen TILP Rechtsanwälte gegen Volkswagen Aktiengesellschaft, 55.

195 **The opinion noted that the EPA**: Kirkland & Ellis LLP, memo to Volkswagen, "Re: Emissions Control and Onboard Diagnostics Compliance Issues—Regulatory Overview," Aug. 6, 2015.

195 **That was the message that Volkswagen**: Göhmann Rechtsanwälte, In Sachen TILP Rechtsanwälte gegen Volkswagen Aktiengesellschaft, 7–8.

195 **That was true of Hyundai-Kia**: United States of America; California Air Resources vs. Hyundai Motor Company; Hyundai Motor America; Kia Motors Corporation; Kia Motors America; Hyundai America Technical Center, Inc., Consent Decree, Nov. 3, 2014, p. 4.

196 **Since 2008, Volkswagen had sworn**: United States of America v. James Robert Liang, grand jury indictment, 2–3.

196 **James Robert Liang, a Volkswagen employee**: United States of America v. James Robert Liang, plea agreement.

196 **After the West Virginia study came out**: United States of America v. James Robert Liang, plea agreement, 6–7.

196 **In early August, Oliver Schmidt**: State of New York v. Volkswagen AG, 47.

197 **Schmidt and Johnson asked to meet:** Interview with Alberto Ayala, October 10, 2016.

197 **The information provided by Volkswagen**: Ibid.

197 **Ayala did not tell anyone at Volkswagen**: E-mail from Alberto Ayala, Oct. 17, 2016.

197 **They promised to do a second recall**: State of New York v. Volkswagen AG, 48–49.

198 **Needless to say, CARB still refused**: Göhmann Rechtsanwälte, In Sachen TILP Rechtsanwälte gegen Volkswagen Aktiengesellschaft, 58.

198 **That day Johnson admitted**: Ayala interview, October 10, 2016.

198 **despite orders from above**: Volkswagen Plea Agreement, Exhibit 2-24.

199 **delete anything related**: Volkswagen Plea Agreement, Exhibit 2-29, 2-30.

199 **forty employees at Volkswagen and Audi**: Ibid.

199 **Executives still thought they could**: Göhmann Rechtsanwälte, In Sachen TILP Rechtsanwälte gegen Volkswagen Aktiengesellschaft, 7. See also Stuart Johnson e-mail to Cornelius Renken (Volkswagen in-house counsel) re: EPA Notice of Enforcement, January 19, 2016.

199 **"they lied through their teeth"**: Ayala interview, October 10, 2016.

199 **Volkswagen later maintained**: Ibid., 53–59.

200 **"In a meeting on 9/3/2015 with"**: Memo to Martin Winterkorn from W. Zimmermann, Sept, 4, 2015.

CHAPTER 18: EMPIRE

201 **Usually the Ballsporthalle**: The author attended the event and recorded it. Translations are his own.

205 **A few days earlier, Müller**: "Porsche-Chef Müller im Interview: Hybrid—

Die neue Porsche-Strategie?," *Auto, Motor und Sport*, 14 Sept. 2015, http://www.auto-motor-und-sport.de/news/porsche-chef-mueller-interview-hybrid-strategie-9968044.html.

205 **Volkswagen was the only major carmaker**: Interview with Mary Nichols, chairwoman of CARB, Oct, 10, 2106. See http://www.pevcollaborative.org/membership.

206 **In China, Volkswagen sold more cars**: "China: The Second Home Market of the Volkswagen Group" slide presentation by Carsten Isensee, executive vice president finance, Volkswagen Group China, Hong Kong, Nov. 24–25 2014, http://www.volkswagenag.com/content/vwcorp/info_center/en/talks_and_presentations/2015/01/China_Pres.bin.html/binarystorageitem/file/2014-11-24+HSBC+Roadshow.pdf.

206 **Sales were down 2.5 percent**: "Marke Volkswagen Pkw verkauft 4,35 Millionen Fahrzeuge per September, Oct. 16, 2015," Volkswagen press release, http://www.volkswagenag.com/content/vwcorp/info_center/de/news/2015/10/Aak_VW_Brand.html.

206 **The press release left no doubt**: "EPA, California Notify Volkswagen of Clean Air Act Violations / Carmaker allegedly used software that circumvents emissions testing for certain air pollutants," EPA press release, Sept. 18, 2015, https://www.epa.gov/newsreleases/epa-california-notify-volkswagen-clean-air-act-violations-carmaker-allegedly-used.

207 **Elizabeth Humstone, a sixty-seven-year-old**: Telephone interview with Humstone, Oct. 17, 2015.

207 **Volkswagen owners reported**: Volkswagen boasted that real world mileage was better than the EPA ratings in a presentation in March 2015 titled "TDI: U.S. Market Success." The presentation cited testing by Consumer Reports as well as an auto blogger who said he got 50 mpg with his Passat.

209 **the price of SCR systems**: Francisco Posada Sanchez, Anup Bandivadekar, and John German, "Estimated Cost of Emission Reduction Technologies for Light-Duty Vehicles," International Council on Clean Transportation, March 2012, http://www.theicct.org/sites/default/files/publications/ICCT_LDVcostsreport_2012.pdf, pp. 60–64

209 **"I'm going through my head"**: Interview with Dan Carder, May 24, 2016.

CHAPTER 19: AFTERMATH

211 **"There was a certain mood"**: Interview with Matthias Wissmann, April 13, 2016.

213 **Volkswagen admitted on September 22**: "Volkswagen AG informiert" (news release) September 22, 2015, https://www.volkswagenag.com/en/news/2015/9/Ad_hoc_US.html.

213 **"As CEO I accept responsibility"**: Volkswagen AG, "Statement by Prof. Dr. Winterkorn," Sept. 23, 2015, http://media.vw.com/release/1070/.

214 **His resignation was "a result"**: Volkswagen AG, "The Volkswagen Group Is Restructuring: Supervisory Board Passes Resolutions for New Organization," Sept. 25, 2015, http://www.volkswagenag.com/content/vwcorp/info_center/en/news/2015/09/organization.html.

214 **The gossip magazine *Bunte* linked**: "Das Aufregende Leben des Neuen VW-Chefs," *Bunte*, Oct. 5, 2015.

214 **In an interview with the *Süddeutsche***: "Porsche Chef Matthias Müller: Wir Haben Verantwortung," *Süddeutsche Zeitung*, Sept. 5, 2015, http://www.sueddeutsche.de/wirtschaft/porsche-chef-matthias-mueller-wir-haben-verantwortung-1.2635475.

215 **"I am definitely not planning"**: Matthias Müller, speech to Volkswagen managers in Leipzig, Oct. 15, 2015.

215 **"I have huge respect"**: Danny Hakim and Jack Ewing, "Matthias Müller, in the Driver's Seat at Volkswagen," *New York Times*, Oct. 1, 2015, http://www.nytimes.com/2015/10/02/business/international/matthias-muller-in-the-drivers-seat-at-volkswagen.html.

216 **"On behalf of our company"**: Michael Horn, testimony before the House Committee on Energy and Commerce, Subcommittee on Oversight and Investigations, Oct. 8, 2015.

216 **"Based on what I know today"**: Matthias Müller, interview with *Frankfurt Allgemeine Zeitung*, Oct. 6, 2015 (print edition only).

216 **exhaustive and independent**: "Erklärung des Präsidiums des Aufsichtsrats der Volkswagen AG zur Sitzung am 30. September 2015" (news release), Oct. 1, 2015, http://www.volkswagenag.com/content/vwcorp/info_center/de/news/2015/10/AR.html.

217 **the EPA issued another notice**: United States Environmental Protection Agency, Notice of Violation, Nov. 2, 2105, https://www.epa.gov/sites/production/files/2015-10/documents/vw-nov-caa-09-18-15.pdf.

217 **But in November sales of Volkswagen**: "Volkswagen of America Reports November Sales" (Volkswagen news release), http://media.vw.com/doc/1684/volkswagen_of_america_reports_november_sales-november_2015_sales_release-1950388522565dbe450fe08.pdf.

218 **"So basically VW is offering"**: E-mail from Elizabeth Humstone, Nov. 9, 2015.

218 **"A lot of things were subordinated"**: Matthias Müller, conference call with analysts and journalists, Oct. 28, 2015.

219 **"We believe that the emergence"**: "Fitch Downgrades Volkswagen to 'BBB+'; Outlook Negative," *Fitch Ratings*, Nov. 9, 2015, https://www.fitchratings.com/site/pr/993669.

220 **"The performance of the dealer"**: Moody's Investors Service, "Credit Opinion: Volkswagen Bank GmbH," March 1, 2016, https://www.moodys.com/research/Moodys-concludes-review-on-European-captive-auto-finance-institutions--PR_262157.

221 **"I felt betrayed,"**: Interview with Mark Winnett, Jan. 19, 2016.

221 **Volkswagen had to pay investors**: Frank Fiedler, chief financial officer of Volkswagen Financial Services, reply to written questions from author, Jan. 22, 2016.

221 **In November, the company said**: Jack Ewing and Jad Mouawad, "VW Cuts Its R.&D. Budget in Face of Costly Emissions Scandal," *New York Times*, Nov. 20, 2015, http://www.nytimes.com/2015/11/21/business/international/volkswagen-emissions-scandal.html.

222 **The last Phaeton**: "Gläserne Manufaktur wird neu ausgerichtet: Schaufenster für Elektromobilität und Digitalisierung entsteht" (Volkswagen news release), March 18, 2016, https://www.volkswagen-media-services.com/detailpage/-/detail/Glserne-Manufaktur-wird-neu-ausgerichtet-Schaufenster-fr-Elektromobilitt-und-Digitalisierung-entsteht/view/3297765/7a5bbec13158edd433c6630f5ac445da?p_p_auth=dbJ9ifiy.

222 **As Volkswagen liked to point out**: "Volkswagen Making Good Progress with Its Investigation, Technical Solutions, and Group Realignment" (Volkswagen news release), Dec. 10, 2015, http://www.volkswagenag.com/content/vwcorp/info_center/en/news/2015/12/VW_PK.html.

CHAPTER 20: JUSTICE

225 **Sally Quillian Yates, a deputy**: Sally Quillan Yates, "Individual Accountability for Corporate Wrongdoing," memo to assistant attorneys general and the director of the Federal Bureau of Investigation, Sept. 9, 2015, https://www.justice.gov/dag/file/769036/download.

225 **"The United States' efforts to learn"**: United States of America vs. Volkswagen AG, U.S. District Court for the Eastern District of Michigan, Jan. 4, 2016, p. 18.

226 **By mid-December 2015, at least**: U.S. Judicial Panel on Multidistrict Litigation, In re: Volkswagen "Clean Diesel" Marketing, Sales Practices, and Products Liability Litigation, MDL No. 2672, Transfer Order, Dec. 8, 2015, p. 1.

226 **Felix Domke, a self-described hacker**: E-mail exchanges with Felix Domke, Feb. to June 2016.

227 **Domke dissected how the engine**: Daniel Lange and Felix Domke, "The Exhaust Emissions Scandal ('Dieselgate')," Chaos Communication Congress, Dec. 27, 2015, https://www.youtube.com/watch?v=xZSU1FPDiao.

228 **"It's completely unrealistic that"**: Interview with Daniel Lange, Feb. 16, 2016.

228 **"We know we deeply disappointed"**: "Matthias Müller: 'The USA Is and Remains a Core Market for the Volkswagen Group'" (Volkswagen news release), Jan. 11, 2016, http://media.vw.com/release/1129/.

228 **"It was a technical problem"**: Sonari Glinton, "'We Didn't Lie,' Volkswagen CEO Says of Emissions Scandal," National Public Radio website, Jan. 11, 2016, http://www.npr.org/sections/thetwo-way/2016/01/11/462682378/we-didnt-lie-volkswagen-ceo-says-of-emissions-scandal.

228 **"I have to apologize for yesterday"**: Ibid.

229 **"Our patience with Volkswagen"**: Danny Hakim and Jack Ewing, "VW Refuses to Give American States Documents in Emissions Inquiries," *New York Times*, Jan. 8, 2016, http://www.nytimes.com/2016/01/09/business/vw-refuses-to-give-us-states-documents-in-emissions-inquiries.html.

230 **a federal appeals court ruled**: Parkcentral Global Hub Limited et al. vs. Porsche Automobile Holdings SE, f/k/a Dr. Ing. H.C. F. Porsche AG, Wendelin Wiedeking, Holger P. Härter, U.S. Court of Appeals for the Second Circuit, Aug. 15, 2014.

230 **the company tapped Francisco**: Volkswagen AG, "Dr. rer. pol. h. c. Francisco Javier Garcia Sanz Member of the Board of Management of Volkswagen AG, with responsibility for 'Procurement'" (curriculum vitae), Dec. 2015, http://www.volkswagenag.com/content/vwcorp/content/en/the_group/senior_management/garcia_sanz.html.

231 **Schmidt . . . told members**: British House of Commons, Transport Select Committee, "Oral evidence: Volkswagen Group emissions violations, HC 495," Jan. 25, 2016, http://data.parliament.uk/writtenevidence/committeeevidence.svc/evidencedocument/transport-committee/volkswagen-group-emissions-violations/oral/27791.html.

232 **Breyer . . . a former aspiring actor**: Kate Galbraith, "Volkswagen Case Gives Judge, Onetime Aspiring Actor, Role of a Lifetime," *New York Times*, April 19, 2016, http://www.nytimes.com/2016/04/20/business/volkswagen-california-judge-charles-breyer.html.

232 **"It is obviously not a whodunit"**: Reporter's Transcript of Case Management Conference, In re: Volkswagen "Clean Diesel" Marketing, Sales Practices, and Products Liability Litigation, U.S. District Court, Northern District of California, Jan. 21, 2016, p. 30.

233 **Mueller kept a replica**: Jack Ewing and Hiroko Tabuchi, "Behind Volkswagen Settlement, Speed and Compromise," *New York Times*, July 15, 2016, http://www.nytimes.com/2016/07/16/business/international/behind-volkswagen-settlement-speed-and-compromise.html.

234 **"Everybody who was involved"**: Interview with Robert Giuffra, June 27, 2016.

235 **"That was a big deal for us"**: Interview with Mary Nichols, Oct. 10, 2016.

235 **On April 21, after negotiating**: Transcript of proceedings, In re: Volkswagen "Clean Diesel" Marketing, Sales Practices, and Products Liability Litigation, U.S. District Court, Northern District of California, April 21, 2016, p. 8.

235 **Breyer replied, "That's perfect"**: Ibid., 8–9.

236 **Volkswagen agreed to pay $2.7 billion**: "Frequently Asked Questions for Beneficiaries to the Volkswagen Mitigation Trust Agreement," EPA news release, July 2016, https://www.epa.gov/sites/production/files/2016-07/documents/faqvwmitigationtrusdtbeneficariesfirstedition0716.pdf.

238 **Härter was convicted in June 2013**: "Porsche ex-finance chief fined for credit fraud," Deutsche Welle, June 4, 2013. http://www.dw.com/en/porsche-ex-finance-chief-fined-for-credit-fraud/a-16857544?

238 **A subordinate was also convicted**: "Geldstrafe für Ex-Porsche-Finanzchef," Frankfurter Allgemeine Zeitung website, June 4, 2014. http://www.faz.net/aktuell/wirtschaft/kreditbetrug-geldstrafe-fuer-ex-porsche-finanzchef-12208349.html.

239 **Executives from Maple Bank said**: In a coda to the case, Maple Bank was in February 2016 closed by German banking regulators amid an investigation into possible tax evasion and money laundering. The questionable transactions were not related to Porsche. See "BaFin Orders Moratorium on Maple Bank GmbH," Federal Financial Supervisory Authority (BaFin) news release, Feb. 8, 2016, http://www.bafin.de/SharedDocs/Veroeffentlichungen/EN/Meldung/2015/meldung_160207_maple_en.html.

240 **The courtroom was crowded**: The author attended the hearing on March 18, 2016.

242 **"The first generation builds"**: Stephan Aust, interview with Ferdinand Piëch, Vox television, broadcast July 17, 2012, https://www.youtube.com/watch?v=O3Tw779LfHM.

242 **the company had hired**: Volkswagen AG, "Christine Hohmann-Dennhardt" (curriculum vitae), http://www.volkswagenag.com/content/vwcorp/content/en/the_group/senior_management/Hohmann-Dennhardt.html.

243 **"They have been rewarded"**: Jack Ewing, "VW Shareholders Vent: 'They Have Been Rewarded for Failure,'" *New York Times*, June 22, 2016, http://www.nytimes.com/2016/06/23/business/international/volkswagen-shareholder-meeting.html.

243 **"Based on what we now know"**: "Matthias Müller: We have Launched the Biggest Change Process in Volkswagen's History" (Volkswagen news

release), June 22, 2016, http://www.volkswagenag.com/content/vwcorp/info_center/en/news/2016/06/HV_2016.html.

243 **the agency had requested**: Ewing, "VW Shareholders Vent."

244 **"Why are they getting so much"**: Interview with Jürgen Franz, Aug. 5, 2016.

244 **But people in Europe were finding**: Jack Ewing, "In the U.S., VW Owners Get Cash. In Europe, They Get Plastic Tubes," *New York Times*, Aug. 15, 2016, http://www.nytimes.com/2016/08/16/business/international/vw-volkswagen-europe-us-lawsuit-settlement.html.

245 **In Brazil, authorities fined Volkswagen**: Ionut Ungureanu, "Volkswagen Fined $13.2 Million by Brazil's Environmental Agency over Amarok Emissions," *autoevolution*, Nov. 15, 2015, http://www.autoevolution.com/news/volkswagen-fined-18-million-by-brazils-environmental-agency-over-amarok-emissions-101967.html.

245 **South Korea also ordered**: Choe Sang-Hunaug, "South Korea Bans Volkswagen from Selling 80 Models in Country," *New York Times*, Aug. 2, 2016, http://www.nytimes.com/2016/08/03/business/international/south-korea-volkswagen-emissions.html.

245 **In Canada, Volkswagen owners**: Christopher Adams, "Motorists Decry Canada's 'Powerless' Response to Volkswagen Scandal as US Nears Settlement," *National Observer*, Aug. 19, 2016, http://www.nationalobserver.com/2016/08/19/news/motorists-decry-canadas-powerless-response-volkswagen-scandal-us-nears-settlement.

246 **Complaints filed by the states of**: State of New York v. Volkswagen AG et al., Maryland Department of the Environment v. Volkswagen AG et al., Commonwealth of Massachusetts v. Volkswagen AG et al.

246 **Winterkorn and a person identified**: Ibid., 24.

247 **"culture that incentivizes cheating"**: Ibid., 68.

247 **James R. Liang . . . engineer**: United States v. James Robert Liang, grand jury indictment, U.S. District Court, Eastern District of Michigan, Southern Division, June 1, 2016.

247 **Under a plea agreement**: James Robert Liang, plea agreement, U.S. District Court, Eastern District of Michigan, Southern Division, August 31, 2016.

248 **Only three years earlier, FBI agents**: Jack Ewing, "Hedge Fund Manager Found and Jailed in Fraud," *New York Times*, March 10, 2013, http://www.nytimes.com/2013/03/11/business/global/hedge-fund-manager-found-and-jailed-in-fraud.html?_r=0.

248 **"The conduct and the facts"**: Interview with Vermont Attorney General William Sorrell, Sept. 14, 2016.

CHAPTER 21: PUNISHMENT

249 **"We are interrupting this broadcast"**: Neo Magazin Royale mit Jan Böhmermann, "Offizielle Drohung der Vereinigten Staaten von Amerika an das Autoland Deutschland," ZDFneo, https://www.youtube.com/watch?v=oMkLrHPYmWw.

249 **a country that uses more energy**: World Bank, "Energy use (kg of oil equivalent per capita)," http://data.worldbank.org/indicator/EG.USE.PCAP.KG.OE.

250 **General Motors agreed to pay**: U.S. Department of Justice, U.S. Attorney Southern District of New York, "General Motors Company—Deferred Prosecution Agreement," Sept. 17, 2015.

250 **linked to the deaths of 124 people**: Danielle Ivory and Bill Vlasic, "$900 Million Penalty for G.M.'s Deadly Defect Leaves Many Cold," *New York Times*, Sept. 17, 2015, http://www.nytimes.com/2015/09/18/business/gm-to-pay-us-900-million-over-ignition-switch-flaw.html?_r=0.

250 **cost GM somewhat over $6 billion**: General Motors Co., U.S Securities and Exchange Commission Form 10-K, 2015, p. 20.

250 **a study by the German government**: Bundesministerium für Verkehr und digitale Infrastruktur, "Bericht der Untersuchungskommission 'Volkswagen,'" April 2016, pp. 90–91.

251 **A British government study**: British Department of Transport, "Vehicle Emissions Testing Programme," April 2016, p. 22.

251 **According to a study published**: Lifang Hou, Kai Zhang, Moira A. Luthin, and Andrea A. Baccarelli, "Public Health Impact and Economic Costs of Volkswagen's Lack of Compliance with the United States' Emission Standards," *International Journal of Environmental Research and Public Health* 13, no. 9 (2016): 891.

252 **a 2010 Jetta diesel sold**: E-mail from Stanley Young, CARB, Oct. 17, 2106.

252 **About 60 people will die**: Jennifer Chu, "Study: Volkswagen's Emissions Cheat to Cause 60 Premature Deaths in U.S.," MIT news office, Oct. 28, 2015, http://news.mit.edu/2015/volkswagen-emissions-cheat-cause-60-premature-deaths-1029.

253 **have fallen only 40 percent**: John German, International Council on Clean Transportation, "Volkswagen's defeat device scandal," presentation to University of Michigan Conference, "Transportation Economics, Energy and the Environment," Oct. 30, 2015, p. 12.

253 **Enforcement of pollution standards**: Rachel Muncrief, John German, and Joe Schultz, "Defeat Devices under the U.S. and EU Passenger Vehicle

Emissions Testing Regulations," International Council on Clean Transportation, March 2016, pp. 4–5.

253 **below twenty degrees Celsius**: Bericht der Untersuchungskommission "Volkswagen," 90.

255 **"there is a lack of guidance"**: Interview with John German, Sept. 2, 2016.

255 **"Whatever mistakes were made"**: Written Testimony of General Motors Chief Executive Officer Mary Barra before the House Committee on Energy and Commerce Subcommittee on Oversight and Investigations, "The GM Ignition Switch Recall: Why Did It Take So Long?," April 1, 2014, http://media.gm.com/media/us/en/gm/news.detail.html/content/Pages/news/us/en/2014/mar/0331-barra-written-testimony.html.

256 **"a swift and robust internal"**: U.S. Department of Justice, "General Motors Company—Deferred Prosecution Agreement," 4.

256 **"So they were even madder"**: Interview with Mary Nichols, Oct. 10, 2016.

257 **increased the number of internal**: Graham Dietz and Nicole Gillespie, "Rebuilding Trust: How Siemens Atoned for Its Sins," *Guardian*, March 26, 2012, https://www.theguardian.com/sustainable-business/recovering-business-trust-siemens.

257 **"Siemens' cooperation, in a word"**: U.S. Department of Justice, "Transcript of Press Conference Announcing Siemens AG and Three Subsidiaries Plead Guilty to Foreign Corrupt Practices Act Violations," Dec. 15, 2008, https://www.justice.gov/archive/opa/pr/2008/December/08-opa-1112.html.

258 **"There are 600,000 hardworking"**: Interview with Klaus Mohrs, April 27, 2016.

CHAPTER 22: FASTER, HIGHER, FARTHER

261 **An investigation by the Los Angeles**: Testimony to the U.S. Senate Committee on Banking Housing and Urban Affairs by Michael N. Feuer, Los Angeles city attorney, Sept. 20, 2016, http://www.banking.senate.gov/public/_cache/files/e5c17a33-d8b0-4e07-8913-a7aaa1ea334c/506BE968E3DBC0673D2DB0B731F45E61.092016-feuer-testimony.pdf.

262 **John Stumpf, chief executive**: Testimony of John Stumpf, chairman and chief executive officer of Wells Fargo & Co., before the U.S. Senate Committee on Banking, Housing, and Urban Affairs, Sept. 20, 2016, http://www.banking.senate.gov/public/_cache/files/18312ce0-5590-4677-b1ab-981b03d1cbbb/3B18AA6E3A96E50C446E2F601B854CF1.092016-stumpf-testimony.pdf.

262 **Stumpf walked away with**: Lucinda Shen, "Here's How Much Wells Fargo CEO John Stumpf Is Getting to Leave the Bank," *Fortune*, Oct. 13, 2016, http://fortune.com/2016/10/13/wells-fargo-ceo-john-stumpfs-career-ends-with-133-million-payday/.

EPILOGUE

266 **"We got played the fool"**: In re: Volkswagen "Clean Diesel" Marketing, Sales Practices, and Products Liability Litigation, transcript of proceedings, Oct. 18, 2016, p. 34.

266 **"Cars are on the road"**: Ibid., 105.

266 **settlement applying to about 80,000 Audi, Porsche, and Volkswagen cars**: Keith Laing , "VW agrees to spend $1B fix or buy back 3-liter diesels," *Detroit News*, Dec. 20, 2016, http://www.detroitnews.com/story/business/autos/foreign/vw-emissions-scandal/2016/12/20/vw-deal/95664236/.

266 **settled with about 105,000 owners in Canada**: Greg Keenan, "Volkswagen Canada strikes $2.1-billion deal with drivers in emissions scandal," *Globe and Mail*, Dec. 19, 2016, http://www.theglobeandmail.com/report-on-business/vw-canada-settles-with-drivers-over-diesel-emissions-scandal/article33361734/.

266 **Bosch, a defendant in the U.S. lawsuits**: Robert Bosch GmbH press release, "Bosch reaches settlement agreement for diesel vehicles in the U.S." http://www.bosch-presse.de/pressportal/de/en/bosch-reaches-settlement-agreement-for-diesel-vehicles-in-the-u-s-87936.html, February 1, 2017.

267 **"We have always had success"**: Dietmar Hawranek and Armin Mahler, "We Will Do Everything to Bring VW Back" (interview with Wolfgang Porsche and Hans Michel Piëch), *Spiegel Online*, Oct. 26, 2016, http://www.spiegel.de/international/germany/vw-and-dieselgate-wolfgang-porsche-and-hans-michel-piech-a-1117984.html.

267 **A few weeks after the interview appeared**: Jack Ewing, "Volkswagen Emissions Scandal Inquiry Widens to Top Levels," *New York Times*, Nov. 6, 2016, http://www.nytimes.com/2016/11/07/business/inquiry-in-emissions-scandal-widens-to-volkswagens-top-levels.html.

267 **"are parting due to differences"**: Volkswagen AG press release, "Dr. Christine Hohmann-Dennhardt to leave the Volkswagen Group Board of Management by mutual agreement—Hiltrud Werner appointed as successor," January 26, 2017.

267 **Incredibly, evidence emerged in 2016**: Jack Ewing, "New Type of Emissions Cheating Software May Lurk in Audis," *New York Times*, Nov. 12, 2016, http://www.nytimes.com/2016/11/13/business/volkswagen-audi-new-emissions-cheating.html.

268 **Agents arrested Oliver Schmidt**: Jack Ewing, Adam Goldman and Hiroko Tabuchi, "Volkswagen Executive's Trip to U.S. Allowed F.B.I. to Pounce," New York Times, January 9, 2017.

268 **Schmidt apparently believed that**: United States of America vs. Oliver Schmidt, Defendant Oliver Schmidt's Motion for Revocation of the Magistrate

Judge's Detention Order, United States District Court Eastern District of Michigan, February 24, 2017, 1.

268 **A judge ordered**: Keith Laing, "Judge denies bail request from VW exec," *Detroit News*, January 12, 2017.

268 **a motion filed by his lawyers**: United States of America vs. Oliver Schmidt, Defendant Oliver Schmidt's Motion for Revocation of the Magistrate Judge's Detention Order, United States District Court Eastern District of Michigan, February 24, 2017, 6–9.

268 **"Mr. Schmidt's participation"**: Ibid., 8.

268 **denied Schmidt's request for bail**: Melissa Burden, "Ex-VW exec accused in emissions scandal denied release," *The Detroit News,* March 16, 2017, http://www.detroitnews.com/story/business/autos/foreign/2017/03/16/volkswagen-executive/99248826/.

269 **"Statement of Facts"**: Volkswagen Plea Agreement, pages 2-1–2-30.

270 **prosecutors in Michigan**: United States v. Richard Dorenkamp, Heinz-Jakob Neusser, Jens Hadler, Bernd Gottweis, Oliver Schmidt, and Jürgen Peter, Second Superseding Indictment, January 11, 2017.

270 **continued to maintain his innocence**: Alison Smale and Jack Ewing, "Ex-Chief of V.W., Testifying in Germany, Stands His Ground on Emissions Deception," *New York Times*, January 19, 2017.

270 **the state attorney's office in Braunschweig**: Staatsanwaltschaft Braunschweig press release, "Zahl der Beschuldigten steigt," January 27, 2018, http://www.staatsanwaltschaften.niedersachsen.de/startseite/staatsanwaltschaften/braunschweig/presseinformationen/zahl-der-beschuldigten-steigt-150570.html.

271 **According to German press reports**: M. Manske and W. Haentjes, "WEGEN BETRUGSVERDACHT Razzia in Winterkorn-Villa!" Bild Zeitung, January 27, 2017.

271 ***Der Spiegel* magazine reported that Piëch**: Dieter Hawranek, "Piëch beschuldigt VW-Aufsichtsräte," *Der Spiegel*, February 8, 2017, http://www.spiegel.de/wirtschaft/unternehmen/volkswagen-ferdinand-piech-beschuldigt-aufsichtsraete-im-dieselskandal-a-1133747.html.

271 **Winterkorn had assured Piëch**: "Piëch belastet Winterkorn vor Staatsanwaltschaft," *Der Spiegel*, February 3, 2017, http://www.spiegel.de/wirtschaft/unternehmen/volkswagen-ferdinand-piech-belastet-martin-winterkorn-vor-staatsanwaltschaft-a-1133024.html.

271 **the Volkswagen supervisory board said it "emphatically repudiates"**: Statement by the Supervisory Board of Volkswagen AG, February 8, 2017.

271 **Winterkorn would not comment**: Felix Dörr, Statement zur Vorabmeldung des Spiegel vom 3. Februar 2017.

272 **Perhaps the explanation was as simple**: Georg Meck, "Piëchs Rache," *Frankfurter Allgemeine Zeitung*, February 12, 2017, http://www.faz.net/aktuell/

wirtschaft/vw-abgasskandal/diesel-skandal-bei-vw-piechs-rache-14873334.html.

272 **he was trying to sell his stake**: Porsche Automobil Holding SE press release, "Mögliche Veränderung der Aktionärsstruktur," March 17, 2017.

272 **"anyone sitting at this podium"**: Matthias Müller, statement made during news conference in Wolfsburg, March 14, 2017.

273 **searched offices belonging to Jones Day**: *Handelsblatt* was the first to report that the searches included Jones Day, and Volkswagen later confirmed it, while protesting vehemently. Jan Keuchel, Martin Murphy, Volker Votsmeier, "Staatsanwälte durchsuchen US-Kanzlei Jones Day," *Handelsblatt,* March 16, 2017; Volkswagen press release, "Statement by Volkswagen AG," March 16, 2017.

273 **Ansgar Rempp, the partner in charge:** E-mail from Ansgar Rempp, March 16, 2017.

273 **authorized searches of offices belonging to Müller and Stadler:** Amtsgericht München, Beschluss, March 7, 2017 (authorizing searches of Audi AG premises).

273 **"which levels of company hierarchy":** Ibid., 8.

AFTERWORD

274 **searches of the carmakers' offices:** European Commission news release, "Antitrust: Commission confirms inspections in the car sector in Germany," October 23, 2017.

274 **The magazine *Der Spiegel* reported:** Frank Dohmen and Dietmar Hawranek, "Absprachen zu Technik, Kosten, Zulieferern: Das geheime Kartell der deutschen Autobauer," *Der Spiegel*, July 21, 2017.

275 **Auto manufacters and suppliers account for:** Commerzbank research note, "German auto industry: are glory days over?" August 9, 2017.

275 **"The so-called diesel crisis":** Bundesministerium der Finanzen, Monatsbericht des BMF, August 2017, 49.

275 **During a visit to Sacramento in 2010:** Testimony of Mary Nichols, March 6, 2017, before the *Untersuchungskommission Volkswagen*, as described in the investigative committee's final report, 355–66. Translated from German to English by the author.

276 **"Never before or since":** Ibid., 356.

276 **"Is that really a good strategy":** Ibid., 356.

276 **she told *Der Spiegel*:** "Spiegel Gespräch: 'Ich bin empört,'" *Der Spiegel*, Issue 36/2017.

277 **In October 2017, London announced:** Mayor of London news release,

"Mayor's new £10 'T-Charge' starts today in central London," October 23, 2017.

277 **The mayor of London:** Mayor of London website, "London's toxic air is a health crisis," https://www.london.gov.uk/sites/default/files/shorthand/clean_air/.

278 **sentenced to 40 months in prison:** U.S. Department of Justice news release, "Volkswagen Engineer Sentenced for His Role in Conspiracy to Cheat U.S. Emissions Tests," August 25, 2017, https://www.justice.gov/usao-edmi/pr/volkswagen-engineer-sentenced-his-role-conspiracy-cheat-us-emissions-tests.

278 **"massive and stunning" fraud:** Bill Vlasic, "Volkswagen Engineer Gets Prison in Diesel Cheating Case," *The New York Times*, August 25, 2017.

279 **admitted feeding false information:** *United States of America vs. Oliver Schmidt*, plea agreement, U.S. District Court Eastern District of Michigan, August 4, 2017.

280 **Pamio told the Munich prosecutors:** Interview with Walter Lechner, August 21, 2017.

281 **"functionality off-cycle":** Volkswagen internal memo, "Status report, 2.0l TDI," May 4, 2007.

282 **optimize emissions "in the cycle":** Volkswagen internal PowerPoint presentation, "Status 2.0l Common Rail (CR) US '07/Exhaust gas emissions," November 7, 2007.

282 **would not report the measures:** Ibid.

282 **not to forward it to anyone else:** Email from Andreas Specht to numerous Volkswagen employees, November 7, 2007.

282 **"do not support the inference":** Email from Volkswagen Communications Department to the author, May 17, 2017.

282 **"implementation of the presented measures was confirmed":** Email from Falko Rudolph to numerous Volkswagen employees, November 9, 2007.

283 **the financial burden of the fraud:** Volkswagen AG, Interim Report 2017 January–September, 1.

283 **lawsuits by individual diesel owners grew to 4,600:** Ibid., 56.

284 **slippage in market share in the European Union:** European Automobile Manufacturers Association news release, "New Passenger Car Registrations European Union," November 16, 2017.

285 **would invest 34 billion euros:** Volkswagen news release, "Volkswagen Group's planning round commits to investments for the future," November 17, 2017.

INDEX